SpringerBriefs in Applied Sciences and Technology

PoliMI SpringerBriefs

AF173702

Series Editors

Barbara Pernici, DEIB, Politecnico di Milano, Milano, Italy

Stefano Della Torre, DABC, Politecnico di Milano, Milano, Italy

Bianca M. Colosimo, DMEC, Politecnico di Milano, Milano, Italy

Tiziano Faravelli, DCHEM, Politecnico di Milano, Milano, Italy

Roberto Paolucci, DICA, Politecnico di Milano, Milano, Italy

Silvia Piardi, Design, Politecnico di Milano, Milano, Italy

Gabriele Pasqui ⓘ, DASTU, Politecnico di Milano, Milano, Italy

Springer, in cooperation with Politecnico di Milano, publishes the PoliMI Springer-Briefs, concise summaries of cutting-edge research and practical applications across a wide spectrum of fields. Featuring compact volumes of 50 to 125 (150 as a maximum) pages, the series covers a range of contents from professional to academic in the following research areas carried out at Politecnico:

- Aerospace Engineering
- Bioengineering
- Electrical Engineering
- Energy and Nuclear Science and Technology
- Environmental and Infrastructure Engineering
- Industrial Chemistry and Chemical Engineering
- Information Technology
- Management, Economics and Industrial Engineering
- Materials Engineering
- Mathematical Models and Methods in Engineering
- Mechanical Engineering
- Structural Seismic and Geotechnical Engineering
- Built Environment and Construction Engineering
- Physics
- Design and Technologies
- Urban Planning, Design, and Policy

http://www.polimi.it

Andrea Giovanni Mainini · Tiziana Poli ·
Alberto Speroni · Matteo Cavaglià ·
Juan Diego Blanco Cadena

Unlocking the Potential of Building Envelopes

Sustainable and People-Centered Approach
to Reduce the Environmental Footprint
of the Built Environment

POLITECNICO
MILANO 1863

Andrea Giovanni Mainini ⓘ
Department ABC—Architecture, Built
Environment and Construction Engineering
(DABC)
Politecnico di Milano
Milan, Italy

Tiziana Poli ⓘ
Department ABC—Architecture, Built
Environment and Construction Engineering
(DABC)
Politecnico di Milano
Milan, Italy

Alberto Speroni ⓘ
Department ABC—Architecture, Built
Environment and Construction Engineering
(DABC)
Politecnico di Milano
Milan, Italy

Matteo Cavaglià ⓘ
Department ABC—Architecture, Built
Environment and Construction Engineering
(DABC)
Politecnico di Milano
Milan, Italy

Juan Diego Blanco Cadena ⓘ
Department ABC—Architecture, Built
Environment and Construction Engineering
(DABC)
Politecnico di Milano
Milan, Italy

ISSN 2191-530X   ISSN 2191-5318 (electronic)
SpringerBriefs in Applied Sciences and Technology
ISSN 2282-2577   ISSN 2282-2585 (electronic)
PoliMI SpringerBriefs
ISBN 978-3-031-75297-1   ISBN 978-3-031-75298-8 (eBook)
https://doi.org/10.1007/978-3-031-75298-8

This Springer imprint is published by the registered company Springer Nature Switzerland AG
The registered company address is: Gewerbestrasse 11, 6330 Cham, Switzerland

If disposing of this product, please recycle the paper.

Preface

In an era where architectural innovation is rapidly evolving, the building envelope is a critical component in shaping the sustainability and efficiency of modern structures. The purpose of this book, *Unlocking the Potential of the Building Envelope*, is to delve deeply into the transformative journey of building envelope design amid the dual forces of digital and ecological transitions. The significance of this evolution cannot be underestimated, as it challenges the integration of cutting-edge technologies and sustainable practices that are essential for the future of architecture. This book is designed to captivate a diverse audience. It will be particularly interesting for architects and building designers who are keen to incorporate advanced technologies and sustainable methodologies into their projects. Researchers and academics in the fields of architecture, engineering, and environmental design will find it a valuable resource, offering insights into the latest trends and innovations. Industry professionals, including those involved in construction, materials science, and building technology, will gain practical knowledge that can be directly applied to their work. Furthermore, students and educators will benefit from the comprehensive overview and detailed case studies, such as technology developers and innovators in the building sector will discover new opportunities for advancement. By showcasing the latest research, technologies, and methodologies, readers will gain a comprehensive view of the field. The book includes practical examples and case studies from the activities of the SeedLab@DABC (www.seed.polimi.it) at the Architecture, Built Environment and Construction Engineering Department (ABC) at Politecnico di Milano, providing real-world applications and valuable insights. These examples highlight the collaborative efforts between academia, industry, and governmental bodies, emphasizing the importance of innovation in building design. These case studies also demonstrate the successful implementation of advanced building envelope designs and analyze the impact of digital tools and ecological considerations on project outcomes. Additionally, the book provides a forward-looking perspective on the future of building envelopes, emphasizing the crucial roles of sustainability and digital integration.

We would like to express our heartfelt gratitude to the following individuals and organizations who have been influential in the creation of this book. Our deepest thanks go to Agenzia nazionale per le nuove tecnologie, l'energia e lo sviluppo sos-tenibile (ENEA) for their invaluable research cooperation since years. In addition, we recognize as crucial the funding provided by the Ministero dello Sviluppo Economico (MSE) and Ministero dell'università e della Ricerca (MIUR) and Regione Lombardia. We sincerely thank the ABC department for their funding, which has enabled the enhancement of research activities, third mission initiatives, and public engagement, allowing us to upgrade the equipment in our experimental laboratory. We are profoundly grateful to Politecnico di Milano for providing an inspiring academic environment that fosters innovation and to all the colleagues, students, and Ph.D. students that are and were part of our journey and that shared their time and efforts at SeedLab@DABC. Among the ones we would like to thank the ones that were directly involved in the activities described in the book: M. Fiori, E. De Angelis, R. Pizzi, V. Sirago, G. Lobaccaro, F. Pittau, T. Pagnacco, A. Luna Navarro, M. Donato, M. Allen, S. Mirzabeigi, B. Khalili Nasr, G. Basso, A. M. Abdelrahman Mohamed, M. El Shemy, V. Casarini, S. Bovi, M. Colombo, C. Meli, A. Canato, G. Caccia, Pa&Co Architecture, the Elisir Project Consortium, and the INCASE Project Consortium.

A special thanks to M. Zinzi, R. Paolini, A. Zani, and E. Casolari for their expert insights, unwavering support, and significant contributions to our research.

Finally, we extend our sincerest appreciation to our relatives, especially Anna T. and Little Ada, Davide, Jacopo, Vittoria, Anna P., Maria, Franca, Massimo, Fabio, Roberta for their endless patience, understanding, and encouragement throughout this journey. Their support has been the propulsion engine in the successful completion of this work.

Milan, Italy   Andrea Giovanni Mainini
August 2024   Tiziana Poli
                                                                             Alberto Speroni
                                                                            Matteo Cavaglià
                                                                  Juan Diego Blanco Cadena

Contents

1   The Evolution of Building Envelope Design in the Digital
    and Ecological Transition 1
    1.1   Introduction 1
    1.2   Pivotal Drivers of Change in Building Envelopes 2
    1.3   Regulatory Framework and Standards in Building Envelope
          Design 11
    1.4   Evolving Sustainability of the Building Envelope: A Paradigm
          Shift 12
    1.5   The Transformative Influence of Digitalization on Building
          Envelope Design 14
    1.6   Current Challenges and Unaddressed Solutions in Building
          Envelopes 16
          1.6.1   AI/Digital Twin for (Challenges) 16
          1.6.2   AI/Digital Twin for (Criticalities) 17
    References 17

2   The Relevance of Performance-Based Design Within
    Early-Stage Design 21
    2.1   Introduction 21
    2.2   Optimization Technologies in the Framework
          of Computational Design 24
    2.3   The Implementation of Performance-Based Design
          in the Computational Design Framework 28
    References 31

3   Decarbonization-Driven Design: Energy-Efficient, Responsive,
    Zero-Emission, Positive, Advanced Materials, Sustainable
    Alternatives for the Building Envelope 33
    3.1   Introduction 33
    3.2   Static Envelopes with a Dynamic Behaviour: 3D Textured
          Material as a Potential for Technology Transfer 35

3.2.1   Static Shading Device Systems and Their Relevance
        in Modern Architecture   35
3.2.2   The Workflow for Performance Assessment
        of the Shading Layer   38
3.2.3   A Speditive Methodology to Acquire Complex
        Geometries: A Focus on 3D Scanning   40
3.2.4   Static 3d Geometries as Shading Devices: An Example
        of the Energy Use Assessment   42
3.3   Responsive Building Envelope and the Use of Smart Materials   46
3.3.1   Thermo-Bi-Materials and Shape Memory Materials   47
3.3.2   Integrated Design Process for Dynamic Envelope
        Components Utilizing Smart Materials   50
3.3.3   Concepts for Shading Device Systems Utilizing Smart
        Materials in Building Façade   51
3.4   Active Systems in Building Envelopes for Enhanced
      Sustainability   54
3.4.1   The Concept: A Shared Device for EV Charging
        Embedded in the Building Envelope   55
3.4.2   The Device   57
References   60

**4   Human-Centric Design: Comfort, Well-Being, and Health
     Cognitive in Building Envelope Design**   63
4.1   Introduction   63
4.1.1   Design Bandwidth Due to Occupants' Diversity   64
4.1.2   Human-Centric Envelope Design, Personalized
        Monitoring, and Operation   66
4.2   Unlocking Building Occupants' Potential for Boosted
      Envelop and Building Design and Performance   67
4.2.1   Specialized Envelope Design for Users   67
4.2.2   Customized Envelope Operation from Monitoring
        Humans   73
4.3   Challenges on Application and Integration for Building
      Operation   76
References   77

**5   Parametric Building Envelope Design and Technology
     Integration**   81
5.1   Introduction   81
5.2   Energy Efficiency and Sustainability   82
5.3   Background   84
5.4   Parametric Building Envelope Shape: From Customized
      Design to Optimized Production   86
5.4.1   Unlocking Innovation: The Parametric Approach
        in Cement-Textile Composite (CTC) Technology   91

5.4.2   Parametric Performance in Building Envelope: Form
        Energy to Well-Being-Driven Approach   93
5.4.3   View Out as a New Comfort Parameter for Occupant
        Well-Being   97
5.5   Unlocking the Parametric Building Envelope   99
References 100

**6   Investigating Decision-Making Frameworks for Early-Stage
    Performance-Based Building Envelope Design** 103
6.1   Introduction 103
6.2   Harnessing Graph Database Technology for Improved
      Data-Driven Building Envelope Design 105
6.3   Methodology for Text-Based Analysis and Validation
      of Dependency Network 111
6.4   Using Graph Databases to Pre-assess the General Ranking
      System of Design Factors 113
6.5   Implementing Weighted Graph Databases for Efficient
      Retrieval of Ranked Information 119
6.6   Trade-Offs Among Performance Domains: A Mapping
      Exercise Using Graph Databases 119
6.7   Assessing the Use of Graph Database Queries Within
      Simulated Design Workflows 123
References 124

7   Unlocking the Future: Reaching a Conclusion 127

List of Figures

Fig. 1.1   The Bauhaus School Building, Deassau, Walter Gropius (1925–1926) (Sludge G, CC BY-SA 2.0 https://creativec ommons.org/licenses/by-sa/2.0, via Wikimedia Commons)   3

Fig. 1.2   Ville Savoy, Poissy Le Corbusier, 1928–1930 (Alessio Antonietti, CC BY-SA 3.0 https://pl.wikipedia.org/wiki/ Plik:Villa_savoye_veduta_.JPG, via Wikimedia Commons)   4

Fig. 1.3   (Left) The structural façade of the Bibliothèque nationale de France in Paris, designed by Dominique Perrault. (Right) The structural façade of Telefónica Madrid, designed by Rafael de La-Hoz Castanys, with façade consultant Arup, and Ignazio Fernandez Solla   5

Fig. 1.4   (Left) Double envelope of Johnson Wax, Arese. Highlighting the external skin with its point-fixing of the panels (Right) The double envelope with horizontal ventilation of Palazzo Regione Lombardia, Milan, designed by Pei Cobb Freed & Partners Architects   6

Fig. 1.5   (Left) Static solar shading system (double skin). The variation depends on the change of boundary conditions but does not change the performance/configuration of the system (Right) Static solar shading system (double skin). The variation depends on the change of boundary conditions but does not change the performance/ configuration of the system. Munch Museum, Oslo (*Source* SeedLab@DABC and P. Cardazzi)   9

Fig. 1.6   Dynamic façade. University of Kolding, Denmark (*Source* S. Juhl, CC0, via Wikimedia Commons—https://com mons.wikimedia.org/wiki/File:Syddansk_universitet.Cam pus_Kolding.Denmark.2014_(40).JPG)   10

Fig. 2.1   Number of publications per year stored in the Scopus
            online database. The search query included all publications
            that used the terms computational design, parametric
            design, algorithmic design, and generative design
            in the title, keywords, or abstract spaces. The increasing
            rate of publications highlights the relevance and interest
            in the computational design framework   26
Fig. 2.2   Individual publication count based on the use
            of the individual terms computational design, parametric
            design, algorithmic design, and generative design
            in the title, keywords, or abstract spaces   27
Fig. 2.3   Adapted from Caetano et al. [32]. Venn diagram
            about the conceptual overlaps between Parametric
            Design, Algorithmic Design, and Generative Design.
            The image adds the concept of Computational Design
            as the enveloping structure to the three sub-domains   27
Fig. 2.4   The picture on the left shows the flowchart for the standard
            practice of an iterative design process. On the right,
            the same flowchart is expanded to explicitly illustrate
            the actions taken to guide the navigation around various
            building design properties   30
Fig. 3.1   Three Dimensional samples: 1. Metal Mesh with hexagonal
            holes, 2. Metal mesh with circular holes, 3. HDPE Plastic
            mesh, 4. Plissed Metal Filter, 5. Coarse Metal Filter, 6.
            Polyester 3D Textiles, 7 Metal Mesh, 8 Metal Grid. The
            others from a technology transfer approach represents
            many possible alternatives that can be further developed
            and investigated   38
Fig. 3.2   Angular solar performance of the samples shown
            in Fig. 3.1. The left graph illustrates the solar transmittance
            performance for incidence angles ranging from 0° to 60°,
            with 15° increments. The right graph displays the relative
            change compared to normal incidence, where red indicates
            an increase and blue indicates a decrease   38
Fig. 3.3   Workflow for the assessment of the angular solar and visual
            transmittance of a generic shading element sample   39
Fig. 3.4   (Left) Photogrammetry reconstruction of a plastic mesh
            with filled holes; (Right) NeRF refinement and clear void
            reconstruction   41
Fig. 3.5   Standard office unit model: External axonometric (left)
            and plan (right) view and dimension. The highlighted blue
            panel (left) and dotted line (right) are the shading elements
            previously presented   43

Fig. 3.6   Primary energy use comparison for Heating, Cooling
             and Lighting for different shading alternatives
             and orientation for the Rome Scenario 44
Fig. 3.7   The concept of the SMA spring actuated system. In
             the green dashed line, the ideal position of the thermal
             box that drives the activation of the system is depicted,
             following the conceptual scheme on the left. On the right,
             an example of activation is shown, demonstrating
             how local shading and partial activation of the device
             can be advantageous. The parts highlighted in green
             indicate the areas where the transition from open to closed
             configuration is focused 52
Fig. 3.8   In the upper part, the four transition states of the chosen
             TMP which is selected as a shading system,
             while in the lower part, the DGP conditions are depicted
             with the system deactivated (left) and activated (right) 53
Fig. 3.9   (Left) The general schema of the integration of the device
             in the building façade and the exploitation of the new
             functionalities. (Right) The usage of private parking spots
             as shared space for EV Mobility charging 56
Fig. 3.10   Standard scenario for electric vehicle charging (left);
             Optimized scenario 1 Taking benefit of the local energy
             production for EV charging (centre); Optimized scenario
             2: Energy consumption delay using the EV as portable
             storage with V2G performances 57
Fig. 3.11   Construction phases and installation of the devices
             in an existing building, taking benefit of the existing
             supporting structure of the ventilated façade 58
Fig. 3.12   Construction phases and installation of the devices in a new
             building and a new façade with a dedicated supporting
             structure 58
Fig. 4.1   Diagram on occupant-building interaction impact
             for the European context (extracted and edited from [3]) 64
Fig. 4.2   Envelope design and foreseen operation compared to actual
             activation or controls based on building occupants'
             perception and sensation [28] 66
Fig. 4.3   Surveying campaign and ray-tracing comparison
             of reflected solar radiation from flat and curved facades
             of tall buildings 68
Fig. 4.4   Details on the **a** surveying campaign layout
             and **b** the differences between scenarios around noon
             for 25/09 69
Fig. 4.5   Test facility schematic plan, and preliminary results
             on the incident solar radiation in occupants 71

Fig. 4.6   Electrochromic window activation patterns in a summer
            month based on the evaluated personalized indoor
            environmental conditions for the hosted occupants.
            Occupants are numbered based on their proximity
            to the glazed facade facing south, and activation patterns
            follow the criteria explained in [48]   72
Fig. 4.7   Resulting indoor operative temperatures at each
            of the occupants' locations, for February (Left) and August
            (right) considering the effects of direct solar radiation
            and of the devised electrochromic operation profiles
            depicted in [48]   72
Fig. 4.8   Schematic description and images of an occupant's
            field of view during the planned surveying campaign,
            considering environmental physiological responses in real,
            immersed, and virtual reality scenarios   75
Fig. 5.1   CTC realization process   92
Fig. 5.2   CTC design phases   93
Fig. 5.3   The workflow defined within the multi-criteria analyses
            on light and view was carried out as part of the SEEDLab
            @DABC experimentation framework   98
Fig. 5.4   Circular multi-parameter optimization approach
            schematization   100
Fig. 6.1   Adapted and expanded from Fitch [1]. While the original
            image from Fitch generally described the role of "walls"
            in building design (effectively implicitly considering
            windows as part of the walls regarding the schema
            compilation), in this context the data has been reframed
            and expanded to the concept of building envelope design   104
Fig. 6.2   The picture displays the basic structure of a graph database,
            namely the interconnection of two nodes via an edge.
            The picture further details the possible state for an edge
            relationship: undirected (above) and directed (below).
            A graph database can be used to store multiple dataset
            and may be also implemented in the BIM environment   105
Fig. 6.3   Structure of the graph database   108
Fig. 6.4   Detailed view of the graph database entries   109
Fig. 6.5   Example of data potentially stored in the third sublevel
            of the database structure   109
Fig. 6.6   The graph database resulting from the description
            of the dependency network of the different design elements
            once related to a housing setting   110
Fig. 6.7   (Left) The correlation-network of all entities directly
            related to "daylight". (Right) sub-network of all entities
            simultaneously related to the transparent envelope property
            of "Dimensions" in the performance domain of "daylight"   111

Fig. 6.8   Workflow of the cross-examination of thematic
            co-occurrence in the scope of transparent envelope design
            and performance evaluation 112
Fig. 6.9   Term matrix used to structure the necessary search queries 113
Fig. 6.10   Term lists are used to structure the necessary search
            queries in the scope of the many transparent envelope
            design properties 114
Fig. 6.11   The figure illustrates the frequency of the most commonly
            used author keywords across each text group, derived
            from search queries in academic databases. These text
            groups are organized based on the keywords: automation,
            component aesthetics, window surroundings, window
            ergonomics, operability, and safety and security 115
Fig. 6.12   The figure illustrates the frequency of the most commonly
            used author keywords across each text group, derived
            from search queries in academic databases. These text
            groups are organized based on the keywords: location,
            dimensions, materials, glass and coating, layering,
            and maintenance 116
Fig. 6.13   (Left) Highlight Sub-network composed by performance
            domains and transparent envelope properties. The
            edges' thickness is proportional to their weights. Based
            on each property connection towards performance
            domains, a weighted degree is presented and ranked.
            (Right) Histogram based on the weighted degree scores.
            The diagram shows how "dimension", "location"
            and "operability" are the most impactful properties
            regarding well-being 120
Fig. 6.14   (Left) Highlight Sub-network composed by performance
            domains and people behaviours. The edges'
            thickness is proportional to their weights. Similarly,
            to the previous image, based on each behaviour connection
            towards performance domains, a weighted degree
            is presented and ranked. (Right) Histogram based
            on the weighted degree scores. The diagram shows
            how "window state control", "home automation settings"
            and "shading state control" are the most impactful
            behaviours regarding well-being 120
Fig. 6.15   (Left) Highlight Sub-network composed by performance
            domains and transparent envelope properties implementing
            the scores derived from Table 5. (Right) Histogram based
            on the weighted degree scores. The diagram illustrates
            how "glass and coating" and "layering" are now considered
            to be of increased importance 121

Fig. 6.16   (Left) Highlight Sub-network composed by performance
             domains and people behaviours implementing the scores
             derived from Table 5. (Right) Histogram based
             on the weighted degree scores. The diagram shows
             how "shading system type" and "shading state control" are
             now considered to be of increased importance 121
Fig. 6.17   The relationship matrix highlights the relational network
             between the analysed performance domains 122
Fig. 6.18   Simulated workflow displaying the possible
             implementation of the graph database inside a typical
             optimization workflow 124

Chapter 1
The Evolution of Building Envelope Design in the Digital and Ecological Transition

Abstract The building envelope is a critical element that shapes a building's overall performance, affecting not just structural stability and durability but also energy efficiency, occupant comfort, and health. A cutting-edge envelope design can dramatically reduce energy consumption for heating, cooling, and lighting while safeguarding against moisture intrusion and preventing mould and structural damage. Beyond its technical functions, the envelope plays a pivotal role in enhancing indoor environmental quality and occupant well-being. Advancements in building envelope technology are at the forefront of sustainability, pushing the boundaries of what's possible in reducing environmental impact and increasing resilience. By exploring the evolution of these technologies and their dynamic interplay with economic, social, and environmental trends, we uncover vast potential for innovation. Moreover, understanding how people interact with these built environments and technologies is vital, as these interactions shape behaviours and significantly influence economic and environmental outcomes. This introductory chapter connects paradigm shifts with historical overviews, showcasing how these interconnected concepts drive sustainable and innovative solutions in building envelope design.

1.1 Introduction

The building envelope, often referred to as the building's skin, functions as a vital physical barrier separating the conditioned interior space from the external environment. This envelope comprises opaque elements, transparent components such as windows and doors, and roof structures. It plays a critical role in determining a building's overall performance by directly influencing essential aspects of functionality, including energy efficiency, thermal comfort, indoor air quality, and structural stability. A well-designed and properly constructed building envelope facilitates effective climate control, reducing energy consumption for heating, cooling, and lighting. Additionally, it prevents moisture penetration, mitigating the risk of mould growth and structural damage. The performance of the building envelope is closely tied to occupant well-being, as it impacts indoor environmental quality and comfort levels. Moreover, advancements in building envelope technology contribute to sustainability objectives by lowering environmental impact and increasing the built environment's overall resilience. Understanding these relationships is crucial for architects, engineers, and construction professionals dedicated to developing high-performing, sustainable buildings.

Tracing the historical evolution of building envelopes reveals the progression of architectural and construction practices over time. Early structures, which utilised basic materials and techniques for shelter and elementary climate control, have undergone significant transformations. Progress in materials, construction methodologies, and architectural design has continually redefined building envelopes from traditional forms to contemporary, high-performance systems. Studying this progression offers valuable insights into how societal needs, technological advancements, and environmental awareness have influenced building envelope design and construction, from their initial development through to maintenance, regeneration, and eventual decommissioning.

1.2   Pivotal Drivers of Change in Building Envelopes

The evolution of the building envelope stands as a fundamental aspect of architectural development, having experienced profound changes driven by various influential factors. The following list highlights the forces that have significantly impacted the building envelope's evolution. These forces span multiple domains, including internal innovations within the construction sector, technological transfers from other industries, environmental challenges related to climate change and sustainability, economic considerations for cost-effectiveness, and social factors concerning the well-being of individuals and communities. This compilation underscores the diverse influences that have shaped the building envelope's development.

Separation of Structural and Functional Elements in Building Envelopes (Evolution of Building Envelope Functionality—Part I)

The initial major evolution of the building envelope is closely associated with the separation of load-bearing and functional elements [1–3]. Traditionally, load-bearing walls were predominantly opaque and bore structural loads, featuring minimal openings to maintain mechanical integrity. Over time, these walls have undergone a process of dematerialisation, evolving into surfaces with varying proportions of solids and voids. This shift has led to the development of load-bearing frames capable of independently supporting dynamic loads.

The need for natural light, motivated by health considerations, along with advancements in concrete technology, has transformed the building envelope into a lighter and more transparent membrane [4]. This change not only affects the physical structure but also significantly redefines the interior spatial experience. These developments are prominently reflected in the modern architectural movement [5].

Modern architectural icons, such as the Bauhaus Dessau building [6], exemplify the principle of dematerialisation with their use of glass curtain walls (Fig. 1.1). These walls emphasise transparency and light penetration, effectively redefining the traditional boundaries between interior and exterior spaces. Similarly, Le Corbusier's Villa Savoye showcases [7] modern principles with its extensive glass facades and open floor plan, blurring the line between inside and outside while providing panoramic views (Fig. 1.2). Frank Lloyd Wright's Fallingwater integrates seamlessly with its natural surroundings through large expanses of glass and cantilevered forms that connect the interior with the forest. This architectural paradigm shift has not only impacted the aesthetics of buildings but also revolutionised the functionality of the built environment.

Transitioning from Opaque to Transparent Envelope (Evolution of Glazing and Facades Systems Technology)

Initially, windows and transparent façade systems were single-glazed, providing limited thermal and structural performance. Over time, a significant shift in building envelope design has emerged, driven by advancements in technology, safety and environmental concerns, aesthetic aspirations, and the need for cost efficiency [8]. Technological progress has evolved from basic single glazing to more sophisticated systems, with each enhancement improving the performance and functionality of transparent components crucial for modern energy efficiency standards. Today's

Fig. 1.1 The Bauhaus School Building, Deassau, Walter Gropius (1925–1926) (Sludge G, CC BY-SA 2.0 https://creativecommons.org/licenses/by-sa/2.0, via Wikimedia Commons)

Fig. 1.2 Ville Savoy, Poissy Le Corbusier, 1928–1930 (Alessio Antonietti, CC BY-SA 3.0 https://pl.wikipedia.org/wiki/Plik:Villa_savoye_veduta_.JPG, via Wikimedia Commons)

emphasis on dry assembly methods within the DfX framework, compared to traditional construction techniques, further drives innovation in transparent and opaque building envelope systems.

Key technological innovations that have marked the evolution of transparent building envelopes include significant advancements in glass products and production processes, framing technologies, and accessories for air and water tightness and insulation. These innovations have greatly enhanced the efficiency of building envelopes. The integration of new and advanced materials and technologies continues to shape the future of architectural design and sustainability.

The following key advances have driven the transition from predominantly opaque to predominantly transparent building envelopes. Technological advancements in transparent components have consistently balanced innovation in product and function, ensuring continuous improvement and cost–benefit balancing throughout the building envelope's life cycle. This ongoing process enhances the sustainability of a system that is inherently less sustainable. The main drivers of innovation include:

- *Float Glass Process (1920s)*: This method involves floating molten glass on a bed of molten tin, producing large, uniform sheets of glass with minimal distortion and high clarity. The commercialisation of float glass significantly improved the affordability and availability of high-quality glass for architectural use (process innovation).

- *Tempered Glass (1930s–1950s)*: The development of heat-treated glass that is stronger and safer for use in buildings (process/product innovation) [9].
- *Laminated Glass (1960s–1970s)*: Introduction of a technology that sandwiches a layer of plastic between two layers of glass for improved safety and security (product/process innovation) [10].
- *Adoption of Silicone Sealants (since the 1960s)*: Silicone sealants were introduced, offering superior flexibility, weather resistance, and durability. These sealants provided better adhesion to glass and framing materials, improving the longevity of glazed units (product and design innovation) [11].
- *Structural Silicone Glazing (SSG) (since the 1960s)*: This technique involves bonding the glass to the building's structural framing with high-strength silicone adhesives, eliminating the need for mechanical fasteners. This method allows for cleaner lines and larger spans of glass, enhancing both aesthetics and performance (product and design innovation) (Fig. 1.3) [11].
- *Insulated Glass Units (IGUs) (since the 1970s)*: Double and triple-glazed windows/façades provide improved thermal insulation and soundproofing, crucial for modern energy efficiency standards (product innovation) [12, 13].
- *Low-Emissivity (Low-E), Reflective, Sun Control/Selective Glasses (since the 1980s)*: These feature a thin metal or metallic oxide coating, optimising energy and lighting performance (process innovation) [14, 15].

Fig. 1.3 (Left) The structural façade of the Bibliothèque nationale de France in Paris, designed by Dominique Perrault. (Right) The structural façade of Telefónica Madrid, designed by Rafael de La-Hoz Castanys, with façade consultant Arup, and Ignazio Fernandez Solla

- *Patterned/Fritted Glass (since the 1990s)*: These treatments allow for the customisation of façades, modulation of opacity levels during the design phase, control of glare, and assurance of privacy (product innovation) (Fig. 1.3).
- *Self-Cleaning Glass (since the 2000s)*: Coated to break down dirt and grime with sunlight and rain (product innovation).
- *Sealant and Gasket Technology Evolution (since the 2000s)*: High-performance, eco-friendly, and smart sealants and gaskets have enhanced the longevity and performance of insulated glass units (IGUs) and glass façades. These advancements reduce thermal bridging and condensation while significantly improving sealing and durability, contributing to long-term performance and sustainability (product, process, and design innovation).
- *Structural Glazing Technology (since the 2000s)*: This technology allows glass to be used as a structural element (design and product innovation).
- *Evolution of Curtain Wall Systems (since the 1980s)*: Transitioning from early curtain walls to stick-built and unitised systems, this advancement has revolutionised façade design with increased complexity and scale (process, design, and product innovation) (Fig. 1.4) [16].
- *Double-Skin Façades (since the 2000s)*: Featuring an outer layer of glass that creates an insulating cavity, these façades improve thermal performance and energy efficiency, making them ideal for environmentally conscious building designs (design/function innovation) (Fig. 1.4) [17, 18].
- *Smart Glass Technologies (since the 2010s)*: Technologies such as electrochromic, thermochromic, and photochromic glass dynamically adjust their properties in

Fig. 1.4 (Left) Double envelope of Johnson Wax, Arese. Highlighting the external skin with its point-fixing of the panels (Right) The double envelope with horizontal ventilation of Palazzo Regione Lombardia, Milan, designed by Pei Cobb Freed & Partners Architects

response to changes in light and temperature, enhancing comfort and reducing energy costs (design and product innovation).

Continuous improvements in sustainable and eco-friendly production methods for glass, sealants, and gaskets include recycling processes and the development of glass with enhanced performance characteristics for energy efficiency and sustainability [19]. Ongoing research focuses on integrating advanced materials and nanotechnologies to further enhance the properties of architectural glass, including better thermal insulation, energy generation, and dynamic control over light and heat transmission.

Technological advancements have transformed structural engineering into façade engineering, enabling more intricate designs and larger scales.

Transition from Opaque Envelope to Super-Insulated System (Evolution of Insulation Technology)

The progression of insulation materials and building envelope systems demonstrates significant advancements in thermal efficiency and building performance. Initially, traditional insulation materials such as mineral wool, fibreglass batts, and cellulose provided moderate thermal resistance for wall cavities. From the 1970s to the 1990s, improved insulation materials like higher-density fibreglass, rigid foam boards (polystyrene, polyisocyanurate), and spray foam were introduced, enhancing thermal performance, moisture resistance, and structural rigidity.

More recently, the development of superinsulation materials has revolutionised the field. Vacuum Insulation Panels (VIPs) [20] feature a porous core in an airtight envelope, creating a vacuum that provides extremely low thermal conductivity, a thin profile, and suitability for space constraints. Aerogel insulation [21, 22], derived from gels with air replacing the liquid, offers high thermal resistance, is lightweight, and has good fire resistance. Phase Change Materials (PCMs) absorb and release thermal energy during melting and freezing, stabilising indoor temperatures, reducing heating and cooling loads, and improving thermal comfort. Additionally, spray foam insulation, which starts as a liquid foam that expands and hardens upon application, fills gaps and creates an airtight seal, boasting high R-values and excellent air-sealing properties.

Furthermore, multilayer reflective insulation, composed of multiple layers of aluminium foil separated by air spaces or insulating materials, reflects radiant heat and enhances thermal performance when combined with other insulation types. Modern building envelope systems now integrate these advanced insulation materials with air barriers, vapour barriers, and structural elements to ensure airtightness, moisture control, and thermal efficiency. Examples include insulated concrete forms (ICFs) and structural insulated panels (SIPs), which collectively represent the cutting edge of building design and construction [23–26].

From Prescriptive Codes and Standards to Performance-Based Design

Traditionally, the building and construction industries followed prescriptive codes and standards, specifying the materials and methods to be used without necessarily ensuring their performance in real conditions. These codes were based on experience and compliance rather than optimisation and innovation. They focused on individual components of the building envelope, such as windows, walls, and roofs, rather than the entire building system and how its components interacted.

However, this approach has been challenged by the emergence of the performance-based approach, which emphasises the performance outcomes of the building as a whole [27–30]. This method shifts the focus from prescriptive methods and materials to achieving specific, measurable performance criteria. Design and construction processes are driven by desired outcomes, such as energy efficiency, thermal comfort, durability, and sustainability. This approach considers the building envelope as an integrated system, evaluating how different components interact and contribute to the overall performance of the building.

The performance-based approach encourages the use of innovative materials, technologies, and methods that might not be allowed under prescriptive codes but meet or exceed performance goals. This allows for more creative and tailored solutions to meet specific building needs. Performance criteria are defined in measurable terms, such as energy usage per square metre, thermal resistance, or air leakage rates. Building performance is verified through testing and monitoring to ensure that the design goals are achieved in practice.

Additionally, this approach aims to reduce environmental impact through improved energy efficiency, safety and security, reduced waste, and the use of sustainable materials, supporting long-term sustainability goals [31].

From Building Envelope Technology Design to Building Envelope Optimization

The introduction of parametric and algorithmic design tools for optimisation enables a highly detailed approach to envelope design, enhancing both precision and efficiency. The building envelope and the overall structure function together as an integrated system of controllable variables within defined limits, all working towards achieving efficiency and optimisation in production [32–34].

From Stand Alone to Integrated System (Evolution of Building Envelope Design/ functionality—Part II)

The integration of envelope systems with HVAC for energy efficiency represents a paradigm shift. Seamlessly incorporating these systems with broader building envelope strategies enhances performance. This holistic approach ensures that elements such as insulation, windows, and façades work synergistically with HVAC systems to optimise thermal comfort, reduce energy consumption, and improve indoor air quality. This contributes to a more cohesive and effective approach to sustainable building practices.

Transition from Static to Dynamic Building Envelope (Evolution of Building Envelope Functionality—Part III)

Static envelopes are building envelopes that, while incorporating advanced materials or technologies, do not change their properties or behaviour in response to environmental conditions (Fig. 1.5). However, performance-based design requires building envelopes to adapt in the short, mid, and long term, fostering innovation

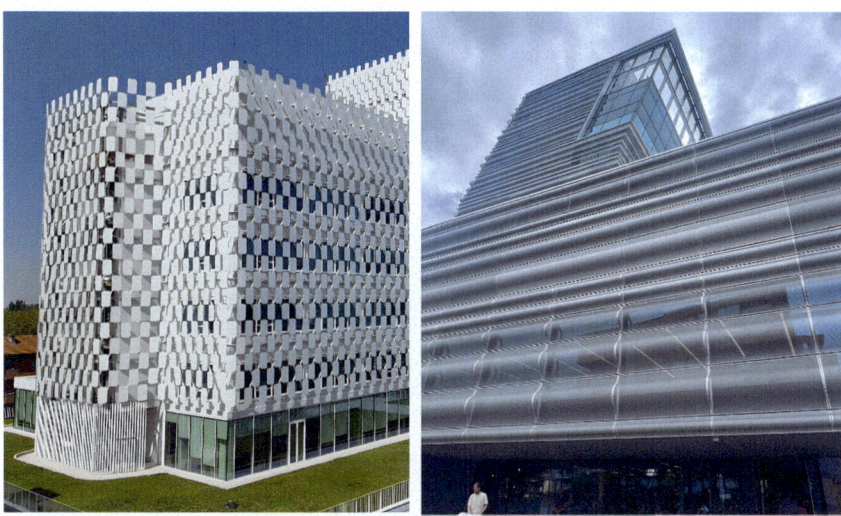

Fig. 1.5 (Left) Static solar shading system (double skin). The variation depends on the change of boundary conditions but does not change the performance/configuration of the system (Right) Static solar shading system (double skin). The variation depends on the change of boundary conditions but does not change the performance/configuration of the system. Munch Museum, Oslo (*Source* SeedLab@DABC and P. Cardazzi)

and transitioning building envelopes from static to dynamic. Dynamic/Adaptive/ Kinetic envelopes [35, 36] (Fig. 1.6) can change their properties or behaviour in response to environmental stimuli such as temperature, light, or humidity, actively optimising indoor comfort and energy efficiency [37].

The Transition from Dynamic to the Responsive Building Envelope (Evolution of Building Envelope Functionality—Part IV)

A responsive envelope is a more advanced concept where the building façade or skin actively changes in real-time in response to environmental stimuli or user preferences. This typically involves the use of sensors, actuators, and control systems to make dynamic adjustments [38].

Transition from Responsive to Smart Building Envelope (Evolution of Building Envelope Functionality—Part V)

Concepts such as the smart envelope underscore a shift towards intelligent and adaptive building envelopes that can dynamically respond to changing conditions or communicate data, integrating the building envelope at a district scale. Moreover, envelopes capable of generating a service, driven by data interactions within the envelope and the broader building-district context, represent a forward-looking fusion of technology and architecture [39, 40]. This transition signifies a fundamental redirection, shifting the focus from individual building-level energy efficiency to a broader perspective at the district scale (from Building Energy Efficiency to District

Fig. 1.6 Dynamic façade. University of Kolding, Denmark (*Source* S. Juhl, CC0, via Wikimedia Commons—https://commons.wikimedia.org/wiki/File:Syddansk_universitet.Campus_Kol ding.Denmark.2014_(40).JPG)

Energy Efficiency). The discourse now encompasses a comprehensive consideration of urban energy systems, infrastructure, and environmental sustainability, emphasizing integrated solutions and holistic approaches to optimise energy performance across entire districts.

A New Paradigm: Cognitive Building Envelope (Evolution of Building Envelope Functionality—Part VI)

Cognitive buildings refer to an advanced approach in architectural design where the building envelope interacts intelligently with the occupants (people-centred design). These buildings can learn from user behaviour and environmental variables to enhance comfort, energy savings, and functionality. From smart energy efficiency to building envelopes informed by user habitude, this concept represents the latest evolution in the functionality of building envelopes, focusing on creating environments that respond dynamically to the needs and behaviours of people within the space via the use of predictive models [41].

Regenerative Envelopes to Promote Well-Being and Health (Evolution of Building Envelope Functionality—Part VII)

The emergence of regenerative envelopes showcases a pioneering perspective on sustainability, as they possess the capability to mitigate vulnerabilities in both the built environment and its occupants [42, 43].

From Introverted to Extroverted Building Envelope (Evolution of Building Envelope Functionality—Part VIII)

The incorporation of interactive glass solutions, informed by behavioural patterns, transforms the building envelope into a communication screen. This evolution enhances its functionality, aligning it more closely with user requirements and environmental considerations. Additionally, leveraging the envelope as a screen introduces innovative marketing opportunities, creating a dynamic platform for engaging visual content and communication.

Biophilic Building Envelope as Part of an Ecosystem (Built Environment)

A biophilic building envelope, as part of an ecosystem within the built environment, represents a paradigm shift in architectural and construction practices. Biophilia refers to the innate human connection with nature, and a biophilic building envelope is designed to incorporate natural elements and patterns into the built environment. This integration of nature into design, often referred to as Nature-Based Solutions (NBS), enhances the connection between humans and their natural surroundings.

1.3 Regulatory Framework and Standards in Building Envelope Design

In the realm of building envelopes, the complexity and breadth of standards and protocols are extensive, encompassing international standards, national and local codes, product-specific regulations, system guidelines, procedural standards, and methodologies for performance measurement and validation. This multifaceted landscape necessitates a holistic and integrative approach, as continuous advancements in building systems reveal the limitations of relying solely on isolated standards. To effectively address the dynamic and complex requirements of modern construction, a cohesive and harmonised multi-domain standard framework is essential. The regulatory framework and standards governing building envelope design are critical to contemporary architectural practices, ensuring compliance with safety, energy efficiency, and sustainability criteria. The primary requirements and performance criteria for building envelopes include structural integrity, airtightness, weather resistance, energy performance, and thermal performance.

Structural integrity demands that the building envelope provide robust support, ensuring stability and resistance to external forces such as wind, seismic activity, and live loads. Airtightness and weather resistance are critical to prevent water ingress, moisture damage, and thermal discomfort, as well as to maintain energy efficiency and indoor air quality by preventing uncontrolled air infiltration.

Energy performance standards and thermal performance requirements are implemented by regulatory bodies to address environmental concerns and enhance building sustainability. Compliance with these standards is essential for reducing energy consumption and mitigating climate impact. Thermal performance standards dictate insulation levels and thermal bridging considerations, optimising energy efficiency and occupant comfort.

Fire safety regulations heavily influence building envelope design, ensuring compliance with standards for fire-resistant materials, cladding systems, and evacuation strategies. Accessibility standards are incorporated into regulatory frameworks to ensure building envelopes cater to individuals with diverse physical abilities, promoting inclusivity in design.

Acoustic performance guidelines are particularly important in urban environments, helping to minimise external noise impact and enhance indoor environmental quality. Increasing concerns about climate change have led to standards addressing the resilience of building envelopes to extreme weather events, safeguarding structures and occupants.

Health, ventilation, and indoor air quality standards significantly influence envelope design, emphasising proper ventilation, material selection, and moisture control to ensure good airflow and prevent pollutant buildup. While performance is predominant, the building envelope must also contribute to architectural aesthetics, aligning with the overall design and visual expectations of the built environment, as exemplified by the new European Bauhaus movement.

Sustainability certification programs, such as LEED (Leadership in Energy and Environmental Design), BREEAM (Building Research Establishment Environmental Assessment Method), or WELL Building Standard, set benchmarks for environmentally responsible building practices, influencing envelope design choices. Durability and longevity are crucial, necessitating materials and construction methods that withstand environmental exposure and the effects of aging.

In addition to traditional domains, one must consider aspects related to POE (Post-Occupancy Evaluation) analysis, along with regulations governing privacy and data security. These are important for defining strategies for the functioning of the envelope and its parts to ensure safety, comfort, health, and energy efficiency.

A key challenge in building envelope design is balancing conflicting requirements by navigating and resolving discrepancies between various codes to achieve a cohesive and compliant design. Additionally, it is crucial to stay updated with evolving standards and seamlessly incorporate new requirements into design practices.

1.4 Evolving Sustainability of the Building Envelope: A Paradigm Shift

The ongoing evolution of the sustainability paradigm within the realm of the building envelope marks a significant departure from conventional design and construction practices. This paradigm shift is driven by an intensified awareness of environmental concerns, energy efficiency imperatives, and the broader goals of creating resilient and eco-friendly structures. The contemporary discourse underscores a heightened environmental consciousness, emphasizing the need for materials and design strategies that minimize ecological impact. Sustainable building practices now prioritize the selection and use of materials based on their environmental footprint, considering every stage from sourcing, and production, to the concluding disposal. Advancements in sustainable materials, particularly recycled and low-impact alternatives, are at the forefront of reshaping the sustainability landscape. These materials not only reduce reliance on virgin resources but often offer superior performance characteristics, including recycled steel, sustainable timber, and novel bio-based materials that provide durability and reduced environmental impact.

This transformation is informed by comprehensive life cycle assessments, which evaluate the environmental impact of the building envelope from material extraction to end-of-life disposal. Life cycle assessments provide a holistic view of sustainability, enabling informed decisions that reduce the overall environmental footprint. By scrutinizing every stage of a building's life cycle, from design and construction to operation and deconstruction, the industry can identify and maximize opportunities to enhance sustainability.

Sustainability now also involves adaptive design strategies that respond dynamically to changing environmental conditions and user needs. Adaptive designs are flexible and resilient, capable of adjusting to variations in climate, occupancy, and use patterns, ensuring long-term sustainability. Techniques such as passive solar design, natural ventilation, and dynamic shading systems exemplify these adaptive strategies, enabling buildings to maintain optimal performance in diverse and evolving conditions.

The evolving paradigm places a heightened emphasis on resilience and climate adaptation, ensuring that building envelopes can withstand and respond to the challenges posed by a changing climate. Resilient design principles focus on creating robust and adaptable structures capable of enduring extreme weather events and long-term climatic shifts. This involves using durable materials, robust construction methods, and innovative design solutions that enhance buildings' ability to cope with and recover from adverse environmental conditions. Integrating innovative technologies like smart sensors and responsive systems further enhances sustainability by enabling real-time adjustments for optimal energy performance. These technologies can monitor and control various aspects of building performance, such as temperature, humidity, and energy usage, contributing to greater energy efficiency and occupant comfort, thereby supporting the overall sustainability goals of modern building envelopes.

Sustainability considerations extend beyond environmental aspects to encompass social and community impact, reflecting a holistic understanding of sustainable building practices. The design and construction of building envelopes now consider their effects on the health, well-being, and social cohesion of communities. This includes ensuring access to natural light, promoting indoor air quality, and creating spaces that foster community interaction and engagement. By addressing these social dimensions, sustainable building practices contribute to creating inclusive and supportive environments. Embracing a paradigm shift in sustainability necessitates interdisciplinary collaboration among architects, engineers, environmental scientists, and policymakers to forge holistic solutions that address the complex challenges of building envelope sustainability. This collaborative approach integrates diverse perspectives and expertise into the design and construction process, leading to more innovative and effective sustainable solutions.

The transformation of the building envelope through the lens of sustainability represents a comprehensive and multifaceted shift in building engineering. It requires a balanced approach that integrates environmental, technological, social, and collaborative elements to create structures that are not only resilient and efficient but also harmonious with their surroundings and beneficial to their occupants. This ongoing evolution highlights the importance of a unified perspective that not only meets current sustainability standards but also pushes the boundaries of sustainable design and construction. Ultimately, this paradigm shift contributes to the creation of resilient, eco-friendly, and socially supportive building environments, setting a new standard for the future of sustainable architecture and construction. By adopting these advanced practices and principles, the building industry can move towards a more sustainable and resilient future, addressing both present and future challenges while promoting a healthier and more equitable built environment.

1.5 The Transformative Influence of Digitalization on Building Envelope Design

The advent of digitalization, marked by the integration of digital technologies, computational tools, and data-driven methodologies, has significantly reshaped the landscape of building envelope design. This transformation begins with the utilization of advanced computational simulations, such as energy modeling and thermal analysis, which empower architects to optimize building envelope performance in terms of energy efficiency, occupant comfort, and sustainability. These simulations provide the ability to make precise predictions and adjustments, ensuring that designs are innovative, environmentally responsible, and cost-effective. By incorporating these advanced techniques, architects can anticipate the behaviour of building envelopes under various conditions and tailor their designs to maximize performance.

In this regard, it is also interesting to note the contribution of Building Information Modeling (BIM), which facilitates a comprehensive and collaborative approach to building envelope design. BIM improves communication and coordination among various stakeholders throughout the lifecycle of a project, from initial conception through construction to maintenance. By integrating all aspects of the design process into a single, cohesive model, BIM ensures that every detail is meticulously planned and executed, reducing errors and enhancing overall project efficiency. This integrated approach helps to align the goals and efforts of all parties involved, fostering a more harmonious and productive working environment.

Pertaining to the optimization of design iterations, generative design algorithms harness the immense computational power available today. These algorithms assist architects in generating innovative and optimized building envelope solutions by evaluating vast amounts of data and providing design options that meet specific criteria and constraints. This process not only accelerates the design phase but also opens up new possibilities for creativity and innovation in building envelope architecture. By leveraging these powerful tools, architects can push the boundaries of traditional design and create structures that are both aesthetically pleasing and highly functional.

Digitalization also plays a crucial role in enabling the creation of responsive environments, where building envelopes dynamically adapt to environmental conditions, user preferences, and real-time data. This adaptability is essential for creating sustainable and energy-efficient buildings that respond to changing weather patterns, occupancy levels, and other variables, thereby enhancing the comfort and well-being of occupants. In this context, advanced digital tools facilitate the exploration and analysis of innovative materials, allowing for a data-driven approach to material selection. By evaluating materials based on performance criteria and sustainability considerations, architects can ensure durability, efficiency, and environmental responsibility in their projects.

The integration of Virtual Reality (VR) and Augmented Reality (AR) technologies further enhances the visualization and communication of building envelope designs. These immersive technologies provide architects, clients, and other stakeholders with a tangible sense of the final product, allowing for better-informed decisions and more effective communication throughout the design process. By enabling stakeholders to experience the design in a virtual environment, these tools help bridge the gap between conceptual ideas and physical reality, fostering a deeper understanding and appreciation of the design.

Moreover, collaborative digital platforms enhance communication and knowledge-sharing among professionals involved in building envelope design. These platforms foster interdisciplinary collaboration, bringing together architects, engineers, contractors, and other stakeholders to work cohesively towards common goals. This collective effort is essential for addressing the complex challenges of modern design and ensuring seamless project integration. Pertaining to this, the concept of digital twin technology represents a significant leap forward. By creating a virtual replica of the physical building envelope, digital twins facilitate real-time monitoring, analysis, and optimization. This technology allows for continuous performance assessment and maintenance, contributing to enhanced building performance, longevity, and sustainability.

These paradigm shifts in design processes and tools signifies a new era of architectural excellence and environmental stewardship. As digital technologies continue to evolve, they will undoubtedly bring further advancements and opportunities for innovation in building envelope design, ensuring that future buildings are more adaptive, resilient, and attuned to the needs of their users and the environment.

1.6 Current Challenges and Unaddressed Solutions in Building Envelopes

The contemporary landscape of building envelope design faces many challenges that demand innovative and comprehensive solutions. These challenges extend across multiple domains, including environmental, technological, and socio-economic aspects. In this regard, noteworthy emphasis is placed on integrating advanced technologies, such as AI and Digital Twins, within building envelopes. While this integration is advocated as a potential solution to multiple issues, it is observed here that it also represents on its own a substantial challenge that the sector must address. Unlocking the full potential of these technologies remains unexplored, particularly in terms of optimizing performance, predictive maintenance, and adaptive responses. Here are some of the Challenges and Criticalities in accordance with the literature about Digital Twin [44–47] and AI in construction [48–51]

1.6.1 AI/Digital Twin for (Challenges)

Climate Resilience

Building envelopes encounter challenges related to climate resilience, necessitating designs that can adapt to increasingly unpredictable and extreme weather conditions. Solutions are sought to mitigate climate-induced stresses on the built environment.

Material Innovation and Lifecycle Assessment

The selection of sustainable and innovative materials for building envelopes poses a persistent challenge. A holistic approach involving lifecycle assessments is required to address the environmental impact of materials from extraction to disposal.

Data drive in a Interdisciplinary approach

Achieving seamless interdisciplinary collaboration among architects, engineers, data scientists, and other stakeholders is an unaddressed challenge. Efficient communication and integration of diverse expertise are essential for holistic and effective building envelope solutions.

Urbanization and Population Growth

The global trend of urbanization and population growth raises challenges in optimizing building envelopes for high-density urban settings. Solutions must address spatial constraints, resource utilization, and the provision of sustainable living spaces.

Adaptive Design Strategies

Developing adaptive design strategies within building envelopes to respond to evolving user needs and preferences remains a challenge. Dynamic, user-centric designs that accommodate changing lifestyles and technologies are sought after.

Regulatory Compliance and Standardization

The evolving landscape of regulations and standards for building envelopes presents an ongoing challenge. Achieving compliance with diverse standards while fostering innovation poses complexities that require careful navigation.

1.6.2 AI/Digital Twin for (Criticalities)

Data Security and Privacy Concerns

The increasing integration of digital technologies introduces concerns regarding data security and privacy within building envelopes. Strategies for safeguarding sensitive information while harnessing the benefits of digitalization present ongoing challenges.

The potential integration of AI and Digital Twin technologies stands as a promising avenue for addressing these challenges. Leveraging these technologies can offer predictive modelling, real-time monitoring, and adaptive solutions that contribute to the optimization and resilience of building envelopes. The intersection of artificial intelligence and digital twins has the potential to revolutionize how building envelopes are conceived, de\signed, and maintained, marking a transformative leap toward more sustainable, efficient, and responsive architectural solutions.

References

1. Roiz A, Garcia Del Valle M (2014) Envelopes: structures at the boundary, pp 2288–2294
2. Kumar G, Raheja G (2016) Design determinants of building envelope for sustainable built environment: a review. Int J Built Environ Sustain 3. https://doi.org/10.11113/ijbes.v3.n2.127
3. Seelow A (2017) Function and form: shifts in modernist architects' design thinking. Arts 6:1. https://doi.org/10.3390/arts6010001
4. Rigone P (2011) Le facciate continue. Maggioli Editore
5. Bachrun AS, Ming TZ, Cinthya A (2019) Building envelope component to control thermal indoor environment in sustainable building: a review. Sinergi 23:79. https://doi.org/10.22441/sinergi.2019.2.001
6. Gropius W (1930) Bauhausbucher 12. Albert Langen Verlag, Dessau
7. Dehghan Y (2013) Design of windows as an external building feature in the works of loos and le corbusier. Archit Res 9:16–22. https://doi.org/10.5923/j.arch.20190901.03
8. Croce S, Poli T (2013) Transparency
9. Soon Ho K, Chang Kyu L (2019) Apparatus and method for manufacturing tempered glass
10. Alkemper Lutz J, Klippe Rüdiger D (2019) Composite glass pane
11. Jang PS, Hong SG, Kim SR (2018) A study on comparison of outdoor wind pressure performance according to outdoor exposure and acceleration deterioration methods of structural sealants applied to curtain wall. J Korea Acad Ind Coop Soc 19:279–287
12. Akram MW, Hasannuzaman M, Cuce E, Cuce PM (2023) Global technological advancement and challenges of glazed window, facade system and vertical greenery-based energy savings in buildings: a comprehensive review. Energy Built Environ 4:206–226. https://doi.org/10.1016/j.enbenv.2021.11.003

13. Jiang S (2006) Advances in glass and optical materials
14. Wondraczek L, Mauro JC (2009) Advancing glasses through fundamental research. J Eur Ceram Soc 29:1227–1234. https://doi.org/10.1016/j.jeurceramsoc.2008.08.006
15. Karasu B, Bereket O, Biryan E, Sanoğlu D (2017) The latest developments in glass science and technology. El-Cezeri Fen ve Mühendislik Dergisi 4:209–233. https://doi.org/10.31202/ecjse.318204
16. Sun Y, Huang Y, Chao L (2019) Novel hidden framing glass curtain wall
17. Lops C, Di Loreto S, Pierantozzi M, Montelpare S (2023) Double-skin façades for building retrofitting and climate change: a case study in central Italy. Appl Sci 13:7629. https://doi.org/10.3390/app13137629
18. Al-Awag EA, Wahab IA (2023) Double-skin facades: advantages, disadvantages, and design considerations. In: techniques and innovation in engineering research, vol 9. B P International (a part of SCIENCEDOMAIN International), pp 146–157
19. Peng K, Xu K (2024) A novel industrial process situation awareness model based on multi-time scale dynamic feature fusion with applications to float glass manufacturing. Can J Chem Eng. https://doi.org/10.1002/cjce.25268
20. Zhou J, Peng Y, Xu J et al (2022) Vacuum insulation arrays as damage-resilient thermal superinsulation materials for energy saving. Joule 6:2358–2371. https://doi.org/10.1016/j.joule.2022.07.015
21. Ma J, Lü Y, Wen H et al (2022) Facile synthesis of super-thermal insulating polyimide aerogel-like films. iScience 25:105641. https://doi.org/10.1016/j.isci.2022.105641
22. Kovács Z, Csík A, Lakatos Á (2023) Thermal stability investigations of different aerogel insulation materials at elevated temperature. Therm Sci Eng Prog 42:101906. https://doi.org/10.1016/j.tsep.2023.101906
23. Bruno R, Bevilacqua P, Ferraro V, Arcuri N (2021) Reflective thermal insulation in non-ventilated air-gaps: experimental and theoretical evaluations on the global heat transfer coefficient. Energy Build 236:110769. https://doi.org/10.1016/j.enbuild.2021.110769
24. Pourghorban A, Kari BM, Asoodeh H (2022) Holistic survey of reflective insulation systems (RISs) in vertical applications in building envelopes under various climatic conditions. Energy 242:122959. https://doi.org/10.1016/j.energy.2021.122959
25. Grubb D (2018) Multi-layered reflective insulation system
26. Ujma A, Umnyakova N (2019) Thermal efficiency of the building envelope with the air layer and reflective coatings. E3S Web Conf 100:00082. https://doi.org/10.1051/e3sconf/201910000082
27. Arowoiya VA, Onososen AO, Moehler RC, Fang Y (2024) Influence of thermal comfort on energy consumption for building occupants: the current state of the art. Buildings 14:1310. https://doi.org/10.3390/buildings14051310
28. Martinez-Paneda M (2023) Towards a holistic performance-based design approach. Struct Eng 101:18–22. https://doi.org/10.56330/RSHC9924
29. Armstrong A, Wright C, Ashe B, Nielsen H (2017) Enabling innovation in building sustainability: Australia's national construction code. Procedia Eng 180:320–330. https://doi.org/10.1016/j.proeng.2017.04.191
30. Shach-Pinsly D, Capeluto IG (2020) From form-based to performance-based codes. Sustainability 12:5657. https://doi.org/10.3390/su12145657
31. Ji Y, Wang W, He Y et al (2023) Performance in generation: an automatic generalizable generative-design-based performance optimization framework for sustainable building design. Energy Build 298:113512. https://doi.org/10.1016/j.enbuild.2023.113512
32. Long LD, Le Toan H, Binh TT, Truong NS (2024) An optimization model for building envelope with energy efficiency objectives based on the BIM design buider-RF-NSGAII algorithm, pp 274–287
33. Chaturvedi PKr, Kumar N, Lamba R, Nirwal V (2023) A Parametric optimization for decision making of building envelope design: a case study of high-rise residential building in Jaipur (India), pp 453–465
34. Gan W, Cao Y, Jiang W et al (2019) Energy-saving design of building envelope based on multiparameter optimization. Math Probl Eng 2019:1–11. https://doi.org/10.1155/2019/5261869

35. Sangeetha (2022) Adapt: an experimental design for the application of thermo-responsive shape memory polymers into building envelopes. Open Access Te Herenga Waka-Victoria University of Wellington
36. Sommese F, Badarnah L, Ausiello G (2022) A critical review of biomimetic building envelopes: towards a bio-adaptive model from nature to architecture. Renew Sustain Energy Rev 169:112850. https://doi.org/10.1016/j.rser.2022.112850
37. Perino M, Serra V (2015) Switching from static to adaptable and dynamic building envelopes: a paradigm shift for the energy efficiency in buildings. J Facade Des Eng 3:143–163. https://doi.org/10.3233/FDE-150039
38. Mohtashami N, Fuchs N, Fotopoulou M et al (2022) State of the art of technologies in adaptive dynamic building envelopes (ADBEs). Energies (Basel) 15:829. https://doi.org/10.3390/en15030829
39. Guillermo B, Jan V, Han V, Irena K (2023) Smart building and district retrofitting for intelligent urban environments. In: Intelligent environments. Elsevier, pp 395–420
40. Tonev K, Kappe S, Krahtova P et al (2018) District-scale data integration by leveraging semantic web technologies: a case in smart cities, pp 289–292
41. (2022) Cognitive buildings. MDPI
42. Kujundzic K, Stamatovic Vuckovic S, Radivojević A (2023) Toward regenerative sustainability: a passive design comfort assessment method of indoor environment. Sustainability 15:840. https://doi.org/10.3390/su15010840
43. Torresin S, Aletta F, Bourdeau E et al (2020) Five questions on the indoor soundscape approach for regenerative buildings. In: Proceedings of 2020 international congress on noise control engineering, INTER-NOISE 2020
44. Omrany H, Al-Obaidi KM, Husain A, Ghaffarianhoseini A (2023) Digital twins in the construction industry: a comprehensive review of current implementations, enabling technologies, and future directions. Sustainability 15:10908. https://doi.org/10.3390/su151410908
45. Zhang Z, Wei Z, Court S et al (2024) A review of digital twin technologies for enhanced sustainability in the construction industry. Buildings 14:1113. https://doi.org/10.3390/buildings14041113
46. Cespedes-Cubides AS, Jradi M (2024) A review of building digital twins to improve energy efficiency in the building operational stage. Energy Inform 7:11. https://doi.org/10.1186/s42162-024-00313-7
47. Tuhaise VV, Tah JHM, Abanda FH (2023) Technologies for digital twin applications in construction. Autom Constr 152:104931. https://doi.org/10.1016/j.autcon.2023.104931
48. Rangasamy V, Yang J-B (2024) The convergence of BIM, AI and IoT: reshaping the future of prefabricated construction. J Build Eng 84:108606. https://doi.org/10.1016/j.jobe.2024.108606
49. Prabhakar V, Belarmin Xavier CS, Abubeker KM (2023) A review on challenges and solutions in the implementation of Ai, IoT and blockchain in construction industry. Mater Today Proc. https://doi.org/10.1016/j.matpr.2023.03.535
50. Regona M, Yigitcanlar T, Xia B, Li RYM (2022) Opportunities and adoption challenges of ai in the construction industry: a PRISMA review. J Open Innov: Technol, Market, Complex 8:45. https://doi.org/10.3390/joitmc8010045
51. Nabizadeh Rafsanjani H, Nabizadeh AH (2023) Towards human-centered artificial intelligence (AI) in architecture, engineering, and construction (AEC) industry. Comput Hum Behav Rep 11:100319. https://doi.org/10.1016/j.chbr.2023.100319

Chapter 2
The Relevance of Performance-Based Design Within Early-Stage Design

Abstract This chapter emphasizes the crucial role of early-stage design in building and building envelope design. It suggests that future challenges in building envelope design will require expanding design strategies to include multi-domain and multi-scale approaches. The complexity of these design processes can be managed with parametric optimization technologies. However, selecting the right parameters for the design problem is often overlooked. The choice of parameters, their evaluation ranking, and the hierarchy of objectives are critical factors that can greatly affect the design outcomes. By focusing on these aspects, this chapter emphasizes the importance of performance-based design strategies to address challenges and optimize building envelopes for better sustainability and functionality.

2.1 Introduction

In architectural design and construction engineering, early-stage design planning serves as a comprehensive term that encapsulates multiple preliminary design processes necessary to outline the overall characteristics of a particular project. Though early-stage design involves multiple operations—from laying the groundwork to define the project's scope, budget, and timeline—it is most commonly associated with the study of conceptual or preliminary design. This process is necessary for envisioning the basic properties of the project design. Essentially, the concept definition carried out in early-stage design is used to plan the project's key characteristics, setting at the same time a precise number of design trajectories that are carried along into the subsequent stages of the project for further development and refinement. For this reason, early-stage design assumes a pivotal role in what can be described as the strategic planning phase of a project, representing the critical juncture where project objectives, limitations, and possibilities are delineated. The challenges outlined at the conclusion of Chap. 1 underscore the intricate technological and conceptual roles of the building envelope in the contemporary built environment, affecting all stages of building design. Addressing these challenges and integrating advancements in knowledge necessitates a holistic and mindful approach. This integration goes beyond the localized implementation of novel technologies within isolated aspects of design. Instead, it requires a comprehensive strategy that accounts for the building envelope's multifaceted influences and interactions throughout the entire design and construction process. In light of this new awareness, early-stage design acquires even greater significance. It becomes not only a crucial driver influencing the proper development of a building but also the initial strategic step in ensuring the innovative, or "unlocking", capacity of the entire design process. Furthermore, it secures the building's legacy within the broader social collective. In this regard, Østergård summaries that when strategic planning within early-stage design establishes a trajectory that constrains the design process to a narrower range of outcomes, it simultaneously impacts the overall design on multiple levels, including the attainable performance of the eventual building [1]. During the first two decades of the twenty-first century, the concept of performance, or more specifically building performance, has acquired an increasingly decisive spot at the forefront of the debate on innovation in the building sector. This development has, in a sense, "supercharged" the concept of building performance with a greater scope of action. Within this perspective, Legge et al. provide a historical definition of building performance, where true performance is defined as the fulfilment of requirements based on the essential needs of a building or complex. This is determined either through scientific assessment, or test results from a reliable evaluation method [2]. While this definition is definitely not being rejected in the modern use of the concept of performance, the perception derived from the complexity involved within the descriptive operation has certainly been redefined. This is because due to the influence of digitalization in building design (see Chap. 1, Sect. 1.4), it has been possible to further perceive an increased awareness about the complex cause-and-effect relationships between various design factors. This has emphasized the tangible impact, and risks, of adapting designs to meet performance requirements, as any changes in design can reverberate across multiple domains, significantly altering the building's characteristics regarding altogether different aspects in complex ways. Li et al. further underlined this by affirming that early-stage decisions have a higher potential to influence building performance compared with the decisions usually made at later design stages [3]. This results from the fact that multiple aspects of building performances are the outcome of intricate interactions among various characteristics of the designed environment. Indeed, architectural design can be regarded as a multi-criteria decision problem [4], in which the designer considers various parameters, requiring a combination of experience, knowledge, skills, and the designer's imagination [5]. The necessity of effectively managing this complexity has prompted a fundamental increased attention towards the performance-based design paradigm. This methodology has become an essential framework for addressing the complex challenges inherent in contemporary architectural and urban planning. Performance-based design entails a comprehensive evaluation of a building's potential impact and functionality across multiple domains, ensuring that each project not only meets rigorous standards but also enhances the synergistic interactions among these dimensions. Oxman defines performance-based design as the strategic use of building performance simulation to modify and refine the geometrical form, aiming to optimize a given design in alignment with specific performance objectives [6]. Indeed, the success of performance-based design relies on the existence of metrics that can adequately address specific needs, and to achieve this, it's essential for these performance standards to be quantifiable and verifiable through reliable testing methods [7]. However, while the strategic actions planned in early-stage design have a great impact over the future building performances, it is hard to properly address these issues due to the inherent uncertainties present at this point of the design process. For this reason, as noted by Schlueter et al., performance simulations are mostly executed after the design stage, and therefore, they are not simultaneously integrated into the design decision-making strategies [8]. The output of early-stage design is typically integrated into performance evaluations after defining a specific design instance. Subsequently, this design instance is studied through performance simulations. Peterson et al. highlight that this design process essentially acts as an iterative procedure where the performance-oriented analysis allows to validate the proposed design solution. However, if this validation is not accomplished, namely the resulting performance is evaluated negatively, the designer is forced to repeat different design iterations until a satisfactory performance is reached [9]. Despite the efficiency of this workflow, it must be noted how building performance analysis tools typically provide deterministic results that evaluate the present design [1], generally leaving to the personal experience of the designer the duty of improving the current solution after solving any relevant issue. While numerous studies address this iterative process to simplify [10] the implementation of performance-based strategies during early-stage design [8], it's important to note how dominating a comprehensive understanding of the impact of multiple performances in early-stage design remains a complex undertaking. This is primarily due to the time-intensive nature of modelling, the high rate of changes within early-stage design, input unpredictability, and the conflicting relationships among different requirements and performances [11]. In fact, despite the effectiveness of performance analysis as a form of design validation, early-stage design trials are usually too coarse to serve as a reliable input for meaningful data. On this note, Picco et al. argued that the extraordinary precision and detailed outputs produced by advanced software environments may be unattainable during early design stages, given the limited information and the substantial uncertainties at that stage of definition [12], while for the same reasons Hygh et al. suggested that too highly precise estimations may not hold significance for the conceptual design phase [13].

Although these statements should by no means be interpreted as negating the importance of performance-based considerations in early-stage planning, they should prompt an examination of whether this phase of the process could benefit from an alternative approach. Taking the previous discussion into consideration, it is possible to see that early-stage design primarily aims to develop strategic decision-making processes. This suggests that highly beneficial assistance could come from performance-oriented tools and methodologies aimed to facilitate such procedures. Ideally, these resources would offer crucial guidance and supplementary information for decision-making, even prior to reaching a sufficiently precise design solution for meaningful validation through standard performance evaluation procedures. In addition, a key aspect to underline within the context of early-stage building design is that the focus should not be limited to the examination and validation of individual performance parameters. Rather, a broader perspective should be adopted to account for the capabilities of decision-making strategies to impact multiple types of building performance domains simultaneously. This is a well-known issue in both the research community and the construction sector. However, while an extensive series of projects and research has certainly refined the concept of multi-objective optimization problems, effectively offering methodologies and tools [14] to tackle the simultaneous resolution of multiple design variables, such approaches often suffer from the same previously highlighted constraints in their application to early-stage design. Additionally, multi-domain performance design inherently requires managing multiple parameters which may possess intricate mutual relationships determined by trade-off, synergies, or even proper antagonisms. For example, Alkhatib et al. flatly state how it is particularly hard to create a construction form that meets both aesthetic and functional criteria, exactly because a simple change to one element—or parameter—may cause an impact on other elements [15]. This mutual dependency is common among many elements of the built environment, and even in the specific scope of the design of a building envelope many examples could be formulated. For instance, while increasing window size enhances natural lighting and subsequently reduces energy use for artificial lighting, it can also rise heat loss or gain, disrupting the building's thermal performance. Other elements such as structural requirements and insulation, or durability and sustainability, can often stand in opposition [16, 17]. This intricate web of relationships can become extremely difficult to navigate. Still, it is exactly this kind of data which may be meaningful to access within the decision-making process of early-stage design. For example, multi-criteria and multi-domain optimization tools are frequently used in the design practice to simultaneously address this intricate network of relationships among building features and achieve an overall equilibrium across multiple performance targets [18–21].

2.2   Optimization Technologies in the Framework of Computational Design

To effectively manage the complexity required for implementing performance-based design, suitable digital environments are essential, and the computational design paradigm provides the necessary methodological framework in most instances. Computational design refers to the use of computer-based algorithms and computational processes to aid in the creation, analysis, and optimization of complex design solutions [22]. By employing computational tools, designers can automate repetitive tasks, integrate data-driven decision-making, and increase the levels of control and efficiency within the development of the design process. Computational design methodologies have exhibited a remarkable rate of development in the past years. While it may not be possible to pinpoint a single contextual factor as the decisive driver for the high rate of adoption and diffusion of computational design paradigms, it is interesting to note the synergy between multiple aspects that collectively it could be said to have contributed to increasing both the interest and the adoption of computational design. Foremost among these factors is the recognition of the role that various academic associations of researchers at different levels and in multiple countries have played in fostering fundamental capabilities. These associations have created valuable networks for the exchange of knowledge in the field. Notable among these are eCAADe (Education and research in Computer Aided Architectural Design in Europe) [23], ACADIA (Association for Computer-Aided Design in Architecture) [24], the CAAD Futures Foundation [25], CAADRIA (Association for Computer-Aided Architectural Design Research in Asia) [26], SIGraDi (Sociedade Ibero-Americana de Gráfica Digital) [27], and ASCAAD (Arab Society for Computer-Aided Architectural Design) [28]. Each of these groups is also responsible for organizing conferences at regular intervals, thereby providing physical opportunities to access the latest advancements in the field. In this regard, research on computational design often references other related labels such as parametric design, algorithmic design, and generative design. The specific conceptual definitions and scope of application for each term have characterized multiple studies in the decade between 2010 and 2020. This was in response to a real need for better taxonomic clarity in the field of computational design. The differences in meaning of each label were often not exhaustively detailed in most studies, with extreme cases where these terms were used interchangeably as synonyms. In this regard, the works of Stavric et al. [29] and Michelle et al. [30] represent an example of optimal milestones along this research trajectory aimed to ensure improved clarity in term use. Computational design can also be considered an umbrella term that groups and encompasses the domains represented by the other definitions [31].

Awareness of the link between computational design and its sub-domains of application can subsequently help to direct critical analysis of the effective trends in research interest in these topics. To grasp the scale of the subject, Fig. 2.1 presents the cumulative publication rate from 1990 to May 2024 for academic publications that cite the terms computational design, parametric design, algorithmic design, and generative design, as stored in the Scopus online database. Additionally, Fig. 2.2 further elaborates on the previous data by separating the use of each major label into distinct queries to highlight the overall number of publications addressing each of the four terms related to computational design practice. This count does not exclude the simultaneous use of different labels in a single product. If a publication simultaneously refers to multiple labels, it is counted within each relevant label group. This ensures that each label category is represented by the entire set of publications using that label.

Notably, the rate of use per each label is not homogeneous across the defined timeframe, further consolidating the non-overlapping nature of these concepts and implying the presence of different characteristics between each term, nevertheless, the rapid increase in publication rate is an apparent indicator of the interest around these topics. An important contribution to define a comprehensive and internally coherent taxonomy about computational design and its sub-domains is the work of

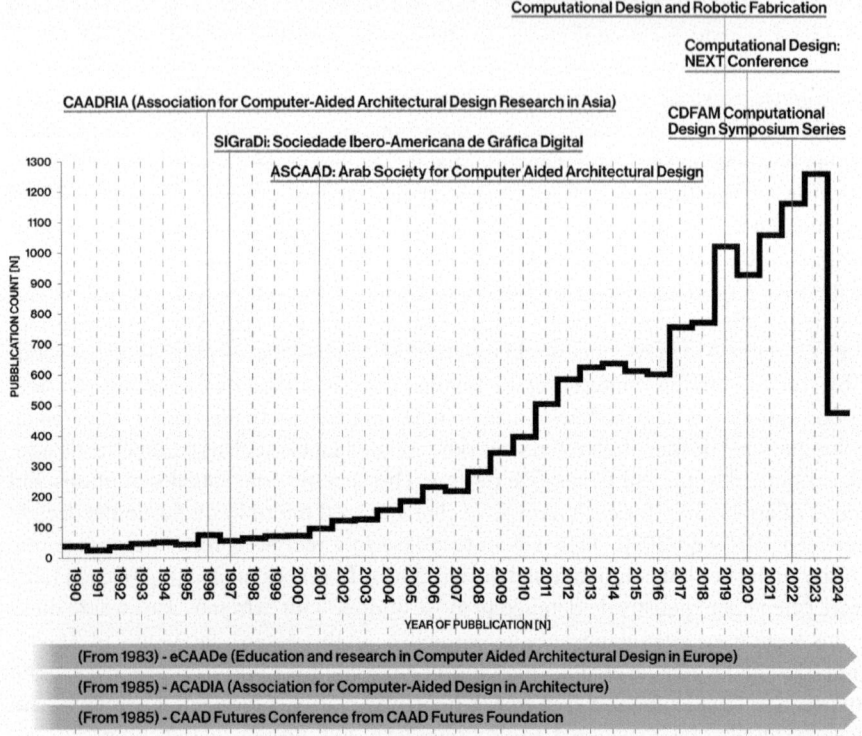

Fig. 2.1 Number of publications per year stored in the Scopus online database. The search query included all publications that used the terms computational design, parametric design, algorithmic design, and generative design in the title, keywords, or abstract spaces. The increasing rate of publications highlights the relevance and interest in the computational design framework

Caetano et al. [32], with Fig. 2.3 representing the overall summary of the authors view on this topic.

As per the author's study, it is possible to define specific characteristics for each of the three sub-domains of computational design. However, these characteristics are not intended to serve as exclusive classification criteria. As illustrated in Fig. 2.3, overlaps between sub-domains do exist, allowing specific design frameworks to border multiple sub-domains simultaneously. However, depending on the framework's goals and methodology, it may still be possible to identify its orientation as being polarized towards a specific sub-domain.

In each sub-domain, Caetano et al. underline the following characteristics:

- *Parametric design* is a modelling framework where the model depends on a set list of parameters that describe it. In parametric design it is emphasized the capacity to affect a design by varying input parameters;

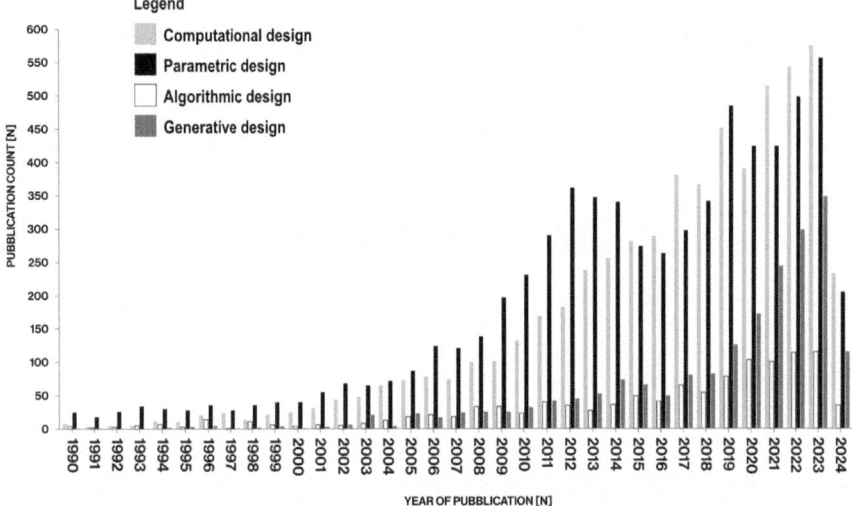

Fig. 2.2 Individual publication count based on the use of the individual terms computational design, parametric design, algorithmic design, and generative design in the title, keywords, or abstract spaces

Fig. 2.3 Adapted from Caetano et al. [32]. Venn diagram about the conceptual overlaps between Parametric Design, Algorithmic Design, and Generative Design. The image adds the concept of Computational Design as the enveloping structure to the three sub-domains

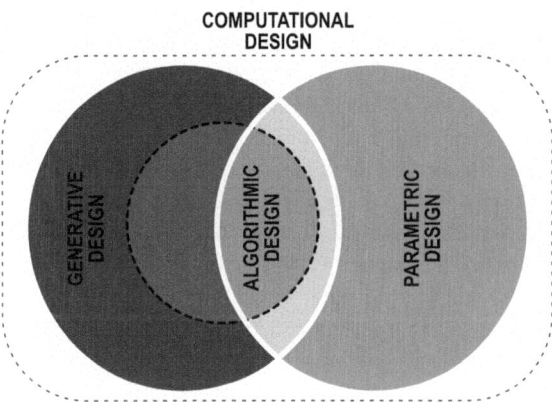

- *Generative design* is a modelling framework where the model depends on data processed within an algorithm that exercises more autonomy than those in parametric design. In this case, the model is not only dependent on input parameters but also on the execution time and method of the generative algorithm. The algorithm may iteratively repeat actions until a specific stop criterion is satisfied, and each iteration may produce notably diverse outputs, making generative design execution less predictable;
- *Algorithmic design* is considered by Caetano et al. as a subset of the generative design framework. The main aspect that identifies an approach as algorithmic is the traceability between the input and output of the algorithmic process. The

authors describe the algorithm within algorithmic design as isomorphic to the model. In essence, algorithmic design streamlines the design process to reach a particular result, rather than expanding the possibilities for unexpected outcomes by increasing the autonomy of the system.

2.3   The Implementation of Performance-Based Design in the Computational Design Framework

Multi-objective optimization procedures intersect with the sub-domains of parametric design and generative design, depending on the level of complexity they present. Rane et al. highlight both parametric and generative design as cutting-edge technologies with significant potential to reshape the architectural design landscape in the near future [33]. Li et al. specifically emphasize the fundamental importance of parametric optimization technology in meeting future building envelope design needs [34]. Building envelope design is expected to face increasingly complex design goals in the near future, extending beyond sustainability issues (see Chap. 1, Sect 1.5). This is because, as Cody observed, true sustainable development cannot compromise the quality attributes of the built environment [35]. Thus, design cannot optimize a single parameter at the expense of others, such as the aesthetic quality of the built environment, without risking the creation of inhuman habitats. Li et al. build upon this observation by proposing that a sustainable building should be conceived as the sum of three aspects: minimal energy consumption, an optimal urban environment, and exceptional spatial quality [34]. This new perspective frames again the building design as a multi-domain optimization problem. However, it is important to note that most parametric optimization platforms focus on providing only the right mathematical models for optimizing the various quantities defined by the users. Consequently, they lack the capability to label or suggest the correct parameters to address for optimizing a given state of building design.

Once again, this is not a simple matter, as it still represents a form of strategic decision-making with the long-lasting capability to impact the entire design process. Addressing the wrong parameters or failing to correctly consider the ramifications of certain parameters on different building aspects can indeed be seen as a potential cause of future project delays and subsequent time or cost overruns.

To clarify, time overrun refers to the duration when a segment of a construction project is finished later than the initially agreed-upon completion date or is not executed as originally planned due to unexpected events or circumstances [36]. Cost overrun, on the other hand, can be defined as the excess of the actual cost over the scheduled budget [37, 38]. While both issues can result from multiple factors, numerous studies have highlighted that the need to frequently alter the design is among the ones with the highest potential to generate both time and cost overruns. Eboh et al. ranked the need for frequent design changes as the 4th factor mostly capable of being responsible for cost overruns from a survey delivered to a sample of 126 professionals in building sector [39], while a literature review published on this topic by Memon et al. placed frequent design changes as the 8th most frequently studied parameter within a sample of 46 other research on this topic [40]. The same study also performed an original survey-based evaluation on the subject, placing the factor labelled as "Poor design and delays in design" as the most critical factor for cost overruns, classifying it as "extremely significant" among a list of 78 other factors [40].

While it could be argued that site-specific factors like soil properties and the economic environment of the construction country should be also considered to properly weigh these considerations, it is nevertheless worthy to acknowledge the literature's repeated emphasis on how proper strategic design planning significantly influences a project's safe execution, as it decreases the risk for future design modifications.

However, regarding the multi-domain parameters dependencies inherent to any building element, Østergård et al. note how it is still not enough to possess an overview about this series of relationships, since also the order used to address different parameters has a huge impact over the overall design trajectory that will be projected [1]. For example, initiating to address privacy issues determined by the configuration of a façade, might prompt the design process to shift towards a different path from the one that may have been selected if daylighting issues would be given priority. The concept of parameters "rank-order" is also used by Heiselberg et al. who proposed the use of sensitivity analysis to determine the specific impact magnitude of different parameters in influencing the design strategy to reach a favourable degree of performance in a specific domain (e.g. energy efficiency) [41]. In this regard, the work of Heiselberg et al. states that not all parameters which may be used to edit a corresponding design, hold equal importance to one another in the framework of achieving specific results.

In conclusion, Fig. 2.4 provides a summary of the reference framework outlined for the iterative approach to performance-based design in the context of early-stage design. Specifically, the image expands on the workflow step where various parameters are selected to be worked upon, presenting a more detailed and organic sequence of sub-steps focused on acquiring comprehensive parameter lists and ranking processes based on their impact on design objectives.

It is proposed here that access to this kind of information during the conceptual stage of a project can effectively assist the "unlocking" of an increased capacity to perform more informed decisions in the proposal for complex design strategies. This workflow primarily seeks to reduce design uncertainty that is inherent in early-stage design by offering valuable information about the potential targets for the impact of design changes, rather than evaluating the specific consequences of each tactical decision.

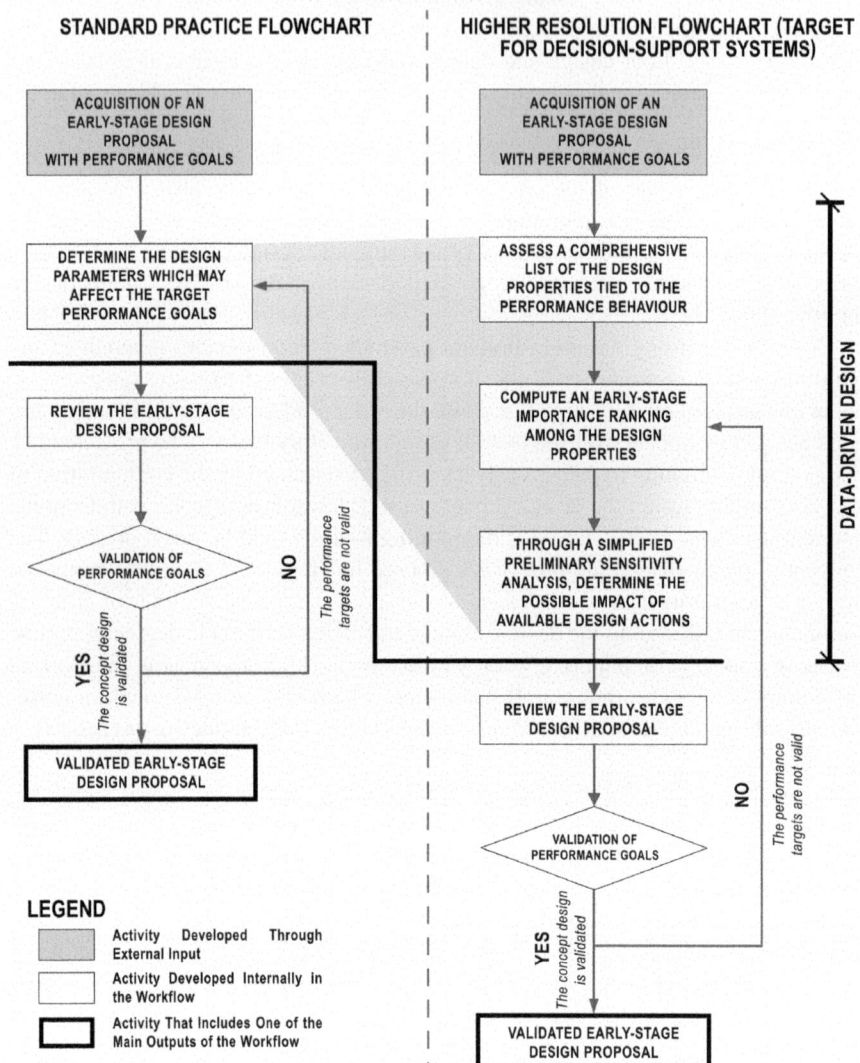

Fig. 2.4 The picture on the left shows the flowchart for the standard practice of an iterative design process. On the right, the same flowchart is expanded to explicitly illustrate the actions taken to guide the navigation around various building design properties

References

1. Østergård T, Jensen RL, Maagaard SE (2017) Early building design: informed decision-making by exploring multidimensional design space using sensitivity analysis. Energy Build 142:8–22. https://doi.org/10.1016/j.enbuild.2017.02.059
2. Legget RF, Hutcheon NB Performance concept in building. In: relation of testing and service performance. ASTM International, 100 Barr Harbor Drive, PO Box C700, West Conshohocken, PA 19428-2959, pp 84–84-12
3. Li J, Bi X, Yang W (2022) Suggestions for solution space exploration in the early stage of architectural design based on a literature review. IOP Conf Ser Earth Environ Sci 1078:012039. https://doi.org/10.1088/1755-1315/1078/1/012039
4. Chan C-S (1990) Cognitive processes in architectural design problem solving. Des Stud 11:60–80. https://doi.org/10.1016/0142-694X(90)90021-4
5. Al-Saggaf A, Nasir H, Hegazy T (2017) Quantifying the impact of architectural design features on building cost and performance in hot weather regions. In: 6th CSCE-CRC international construction specialty conference 2017, pp 62 1–62 10
6. Oxman R (2008) Performance-based design: current practices and research issues. Int J Archit Comput 6:1–17. https://doi.org/10.1260/147807708784640090
7. Ankrah N, Proverbs D (2005) A framework for measuring construction project performance: overcoming key challenges of performance measurement. In: Association of researchers in construction management, ARCOM 2005—proceedings of the 21st annual conference, pp 959–969
8. Schlueter A, Thesseling F (2009) Building information model based energy/exergy performance assessment in early design stages. Autom Constr 18:153–163. https://doi.org/10.1016/j.autcon.2008.07.003
9. Petersen S, Svendsen S (2010) Method and simulation program informed decisions in the early stages of building design. Energy Build 42:1113–1119. https://doi.org/10.1016/j.enbuild.2010.02.002
10. Urban B, Glicksman L (2007) A rapid building energy model and interface for non-technical users. In: Oak ridge national laboratory, buildings conference X, ASHRAE, pp 1–10
11. Østergård T, Jensen RL, Maagaard SE (2016) Building simulations supporting decision making in early design—a review. Renew Sustain Energy Rev 61:187–201. https://doi.org/10.1016/j.rser.2016.03.045
12. Picco M, Lollini R, Marengo M (2014) Towards energy performance evaluation in early stage building design: a simplification methodology for commercial building models. Energy Build 76:497–505. https://doi.org/10.1016/j.enbuild.2014.03.016
13. Hygh JS, DeCarolis JF, Hill DB, Ranji Ranjithan S (2012) Multivariate regression as an energy assessment tool in early building design. Build Environ 57:165–175. https://doi.org/10.1016/j.buildenv.2012.04.021
14. Hamdy M, Nguyen A-T, Hensen JLM (2016) A performance comparison of multi-objective optimization algorithms for solving nearly-zero-energy-building design problems. Energy Build 121:57–71. https://doi.org/10.1016/j.enbuild.2016.03.035
15. Alkhatib FH, Kasim N, Goh WI, Al-masoodi AHH (2022) Performance-driven evaluation and parametrical design approach for sustainable complex-tall building design at conceptual stage. IOP Conf Ser Earth Environ Sci 1022:012047. https://doi.org/10.1088/1755-1315/1022/1/012047
16. Lamperti Tornaghi M, Loli A, Negro P (2018) Balanced evaluation of structural and environmental performances in building design. Buildings 8:52. https://doi.org/10.3390/buildings8040052
17. Long W-J, Lin C, Tan X-W et al (2020) Structural applications of thermal insulation alkali activated materials WITH reduced graphene oxide. Materials 13:1052. https://doi.org/10.3390/ma13051052
18. Galapagos RD. https://www.grasshopper3d.com/group/galapagos. Accessed 17 Sep 2023

19. OCTOPUS. https://www.food4rhino.com/en/app/octopus. Accessed 17 Sep 2023
20. OPOSSUM—Optimization solver with surrogate models. https://www.food4rhino.com/en/app/opossum-optimization-solver-surrogate-models. Accessed 17 Sep 2023
21. Optimo. https://github.com/mrahmaniasl/Optimo/wiki/0_-Home. Accessed 17 Sep 2023
22. Çağdaş G, Colakoglu B (2009) Computation: the new realm of architectural design
23. ecaade. https://ecaade.org/. Accessed 31 May 2024
24. acadia. http://www.acadia.org/. Accessed 31 May 2024
25. CAAD Futures Foundation. https://sites.google.com/unicamp.br/caadfutures/. Accessed 31 May 2024
26. caadria. https://caadria.org/. Accessed 31 May 2024
27. sigradi. https://sigradi.org/pt/inicio_pt/. Accessed 31 May 2024
28. ascaad. https://ascaad.org/. Accessed 31 May 2024
29. Stavric M, Marina O (2012) Application of generative algorithms in architectural design. Adv Math Comput Methods 37
30. Michelle B, Gemilang MP (2022) Bibliometric analysis of generative design, algorithmic design, and parametric design in architecture. J Artif Intell Archit 1:30–40. https://doi.org/10.24002/jarina.v1i1.4921
31. (2023) Examining the differences between computational, parametric, and generative design. https://www.novatr.com/blog/difference-between-computational-parametric-and-generative-design. Accessed 17 May 2024
32. Caetano I, Santos L, Leitão A (2020) Computational design in architecture: defining parametric, generative, and algorithmic design. Front Archit Res 9:287–300. https://doi.org/10.1016/j.foar.2019.12.008
33. Rane N, Choudhary S, Rane J (2023) Leading-edge technologies for architectural design: a comprehensive review. SSRN Electron J. https://doi.org/10.2139/ssrn.4637891
34. Li S, Liu L, Peng C (2020) A review of performance-oriented architectural design and optimization in the context of sustainability: dividends and challenges. Sustainability 12:1427. https://doi.org/10.3390/su12041427
35. Cody B (2017) Form follows energy. De Gruyter
36. Bramble BB, Callahan MT (2010) Construction delay claims
37. Zhu K, Lin L (2004) A stage—by—stage factor control frame work for cost estimation of construction projects. In: Owners driving innovation international conference
38. Kavuma A, Ock J, Jang H (2019) Factors influencing time and cost overruns on freeform construction projects. KSCE J Civ Eng 23:1442–1450. https://doi.org/10.1007/s12205-019-0447-x
39. Eboh EE, Egolum CC, Ezeokoli FO, Onyia CI (2019) Analysis of factors responsible for project cost variation in Enugu, Nigeria. J Sci Res Rep 1–8. https://doi.org/10.9734/jsrr/2019/v25i3-430180
40. Memon AH, Rahman IA, Azis AAA (2011) Preliminary study on causative factors leading to construction cost overrun. Int J Sustain Constr Eng Technol 2:57–71
41. Heiselberg P, Brohus H, Hesselholt A et al (2009) Application of sensitivity analysis in design of sustainable buildings. Renew Energy 34:2030–2036. https://doi.org/10.1016/j.renene.2009.02.016

Chapter 3
Decarbonization-Driven Design: Energy-Efficient, Responsive, Zero-Emission, Positive, Advanced Materials, Sustainable Alternatives for the Building Envelope

Abstract In a world increasingly aware of climate challenges, decarbonization-driven design has become a crucial priority in architecture and construction engineering. This chapter explores the advanced technologies and materials that enable the creation of energy-efficient, responsive, zero-emission, and energy-positive building envelopes. It emphasizes the importance of sustainability as the cornerstone of innovative design approaches, redefining the building envelope as a dynamic and multifunctional system. By adopting a multi-scale approach, the analysis spans from the facade level to individual elements and products, optimizing energy performance and flow control. Key topics include energy efficient envelope components, shading devices, and the principles of responsive and cognitive building envelopes. Through careful analysis and technology transfer, the chapter highlights the role of advanced materials in achieving decarbonization goals, demonstrating how these materials can expand functional possibilities and unlocking new perspectives on sustainability and efficiency.

3.1 Introduction

In a world increasingly aware of climate challenges, decarbonization-driven design emerges as an essential priority for architecture and construction engineering. This chapter aims to explore the advanced technologies and materials that enable the creation of energy-efficient, responsive, zero-emission, and energy-positive building envelopes. Sustainability thus becomes the cornerstone of an innovative approach that redefines the building envelope, not merely as a physical barrier, but as a dynamic and multifunctional system.

The search for simple yet advanced technologies is fundamental for progress in building envelope design. Innovation must proceed from the production phase to prototyping, through careful analysis of emerging technologies. This iterative process allows for the identification and implementation of solutions that enhance efficiency and reduce environmental impact. Technology transfer plays a crucial role, enabling the adaptation and application of new scientific and technical discoveries to the construction context.

Envelope innovation requires a scale leap, from the macro-scale to the micro-scale, analysing the impact at the level of facade, element, and product. This multi-scale approach allows for the optimization of energy performance and flow control, both inward and outward, creating an integrated network of efficiency. A well-designed envelope can manage energy, control flows, and ensure high performance with managed complexity perceived as low.

The choice of materials and functions of the envelope is interdependent: materials influence functions and vice versa. This chapter explores how the selection of advanced materials can expand the functional possibilities of building envelopes, enhancing sustainability and efficiency. Control of energy flows and environmental conditions thus becomes a fundamental component of design, with the envelope acting as the "skin" of the building, dynamically interacting with the external and internal environment.

The building envelope is not just a passive element, but an active system that produces, consumes, shields, and distributes energy. This managed complexity, if properly designed and integrated, can achieve performances comparable to more expensive systems, ensuring sustainability and efficiency at reduced costs. The facade, the element, and the product integrate into a holistic system that offers new perspectives for sustainability and the decarbonisation of the built environment.

The objective of this chapter is to provide, or better, unlock a new perspective on sustainability and the decarbonisation of buildings, exploring how the envelope can act as a dynamic mediator between the inside and outside. It analyses both the application of known materials and technologies in innovative ways and the introduction of new materials and technologies in established applications. This multi-domain approach allows for achieving targets of energy efficiency, responsiveness, energy positivity, and zero emissions, also investigating the concepts of envelope servitization for a sustainable future.

Decarbonisation-driven design represents a paradigmatic shift in how we design and construct buildings. Through technological innovation and the integration of advanced materials, we can create building envelopes that not only reduce environmental impact but also improve the quality of life for occupants, laying the groundwork for a more sustainable and resilient future.

3.2 Static Envelopes with a Dynamic Behaviour: 3D Textured Material as a Potential for Technology Transfer

3.2.1 Static Shading Device Systems and Their Relevance in Modern Architecture

Solar protection devices are essential for improving building energy efficiency, as their effectiveness in providing shade depends on their optical properties and overall performance [1, 2]. This becomes particularly important in modern architecture, where there is extensive use of transparent materials.

Despite the cost-effectiveness of fixed shading devices, their performance can be limited by variable weather conditions. Simulation results suggest that the properties and types of shading systems should vary according to orientation [3]. Therefore, detailed analysis during the early design stages is crucial to optimize building envelope performance.

Shading not only influences daylighting but also impacts the quality of indoor lighting and energy consumption [4]. To address this, alternative approaches are proposed to balance thermal and daylighting performance [5], bearing in mind that higher system complexity often correlates with increased costs [6]. Typically, standard and manually operated shading systems primarily manage glare, based on anticipated sun conditions, with users rarely adjusting these settings, preferring indoor lighting as needed [7].

Despite the high-quality optical performance evaluations obtained, these latest procedures are time-consuming and costly, thus poorly suited to rapid broad-spectrum screening.

While simulation-based assessments are common, they have limitations, particularly concerning complex geometries and reflectance properties [8]. Accurate determination of optical and solar properties is essential for 3D systems, necessitating a reliable dataset and preliminary material-level analysis before architectural and performance assessments [8].

While fixed shading devices effectively reduce cooling loads in summer, they may increase heating demands in winter [9]. Various approaches, including simplified models and geometries, have been used to assess the thermal and optical performance of shading systems [10]. For instance, studies have focused on perforated screens, metal meshes, and grids, analysing their effectiveness in reducing cooling demands while considering lighting and heating requirements [11, 12].

Despite several solutions already available on the market, such data are seldom available, since commercial instruments, such as standard spectrophotometers, are not suitable to provide accurate measurements. The implementation of a reliable dataset and a preliminary analysis at the material level is necessary to lay the conditions for the successive assessment of the technology by architectural design, as well as energy and lighting performance analyses at the building level.

Metal materials have been employed in many applications thanks to their durability, resistance, and pleasant architectural appearance [13]. Several materials (fabric, metal, plastic, and so on) and geometries to create interesting patterns have been analysed so far, and automatic controls have been implemented to improve the efficiency of the system. Fixed shading devices are widely spread thanks to their cost of production and maintenance, and the unrequired specialized skills for the installation [14]. On the other hand, controllable shading devices allow for the adjustment of the amount of light and solar radiation entering the room in response to weather conditions and the position of the sun [15].

The performance of the solar control system is influenced not only by geometry and activation modes but also by the optical-radiative properties of the material used [16].

Designers might choose from a huge variety of possibilities to find an optimal design specific to the location. The importance of creating rules depending on the location makes the design even more connected to the climate and the latitude. These criteria could help realize a resilient and sustainable building, decreasing the energy demand for heating and cooling while guaranteeing a proper level of natural light.

The effectiveness of solar control systems goes beyond just their geometry and activation methods. Understanding the optical and radiative properties of the materials involved is equally crucial. This is where the current challenges lie, amidst the ongoing exploration of new solutions that push the boundaries of technological and formal experimentation within the component. Challenges arise from the scarcity of data on material radiative properties, or the complexity introduced by miniaturizing system geometries, making virtual modelling and subsequent simulations resource-intensive.

While established systems like venetian blinds, textile elements, and louvers showcase consistent performance, the introduction of new materials—particularly those with intricate variable and 3D geometries—adds layers of complexity. In addition, some alternative presents the capability to adapt to the complex geometries of the building envelope (surfaces with single or double curvature). Performance evaluation becomes especially complex due to angular dependencies both because of the material and the building surfaces. Moreover, when intricate geometries are paired with reflective materials, the evaluation process becomes even more critical.

The annual performance of a surface able to shade solar radiation depends on the overall performance of the constituting elements.

The optical-radiative properties of the individual element necessary for determining the system's performance are as follows:

- Light transmittance (τ_v)
- Light reflectance (ρ_v)
- Solar transmittance (τ_e)
- Solar reflectance (τ_e)

It also depends on composite performance indicators such as the ratio between energy reflection and transmission; between light reflection and transmission; and between light transmission and energy transmission (selectivity); Determining composite indicators allows for weighing the system's behaviour against multiple environmental and energy performance indicators such as thermal comfort, visual comfort, and control of solar gains.

A technology transfer process aimed at unlocking innovative shading systems must involve the application of the following perspectives:

1. *Lightweight*—a lightweight element with high mechanical resistance (minimal deformation under wind action) results in lower stress (lower load) on facade substructures; it also allows for easy handling during installation and maintenance. In the context of movable sunscreens, a lightweight element requires less energy to be operated and a more streamlined design of movement mechanisms.
2. *Complex bi and three-dimensional geometry of the constituent element*—the angular dependence of luminous and energy flux and its variability are determined by the material and its orientation during installation rather than the geometric configuration of the shading system. These are generally materials with three-dimensional geometry.
3. *Miniaturization*—highly performing, continuous, lightweight, and thin elements require the use of miniaturized programmed geometry materials and/or structured surface treatments (screen printing and/or nanostructures). This allows the transfer of macroscopic morphological characteristics into treatments and spreads at the microscopic level.
4. *Light diffusion*—diffusing materials allow for achieving indoor environments with homogeneous illumination levels and contribute to reducing glare phenomena.
5. *Transparency*—the optical characteristics of the materials must not alter the perception (distortion of shape and colour) of the external environment or *Selectivity*

Figure 3.1 presents some materials identified as suitable alternatives for a static shading device system. Figure 3.2 illustrates the angular solar performance of the samples, measured on a vertical plane perpendicular to the samples, with the actual arrangement shown in the photos. Solar transmittance values are measured between 0° (normal incidence) and 60°, in 15° increments, using a large integrating sphere optical bench, as described in [11]. The performance curves quickly demonstrate the theoretical effectiveness of these materials and highlight the importance of proper placement/arrangement on the facade to maximize the shading effect. For some materials, incorrect positioning could even lead to increased transparency.

For example, samples 1 and 2 would exhibit significant angular selectivity if mounted on the facade with a 180° rotation from the arrangement shown in the figure. Other samples, such as 4 and 5, due to their pleating and the combined effect of transparency and mutual shading of the elements that make up their texture, can be favourably used in cases where seasonal variability in performance is required. All samples, however, exhibit characteristic angular performance.

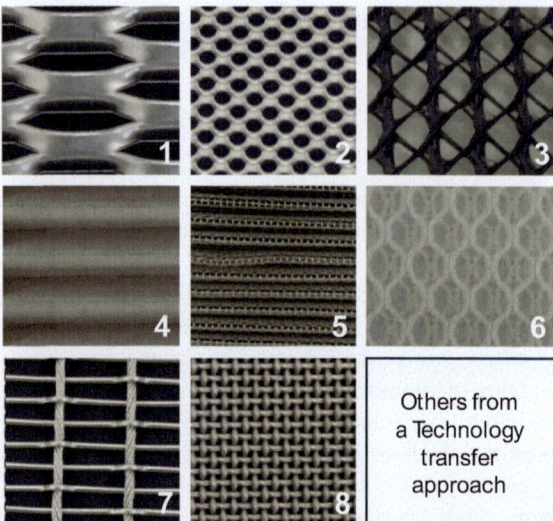

Fig. 3.1 Three Dimensional samples: 1. Metal Mesh with hexagonal holes, 2. Metal mesh with circular holes, 3. HDPE Plastic mesh, 4. Plissed Metal Filter, 5. Coarse Metal Filter, 6. Polyester 3D Textiles, 7 Metal Mesh, 8 Metal Grid. The others from a technology transfer approach represents many possible alternatives that can be further developed and investigated

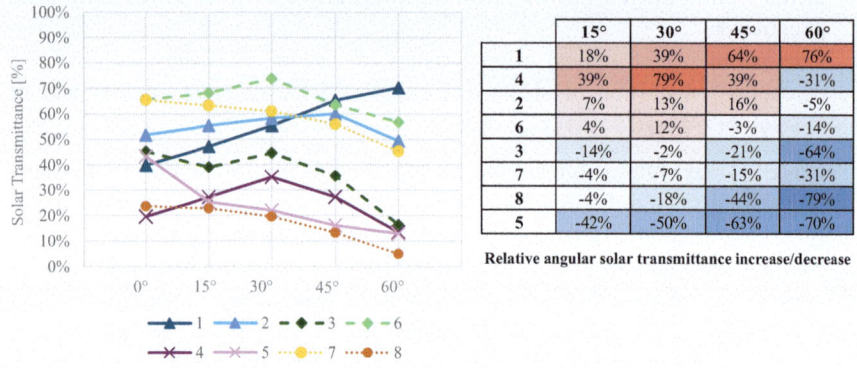

Fig. 3.2 Angular solar performance of the samples shown in Fig. 3.1. The left graph illustrates the solar transmittance performance for incidence angles ranging from 0° to 60°, with 15° increments. The right graph displays the relative change compared to normal incidence, where red indicates an increase and blue indicates a decrease

3.2.2 The Workflow for Performance Assessment of the Shading Layer

In the context of technology transfer, many necessary properties typical of solar shading systems, such as solar and light transmission, are often unknown and require

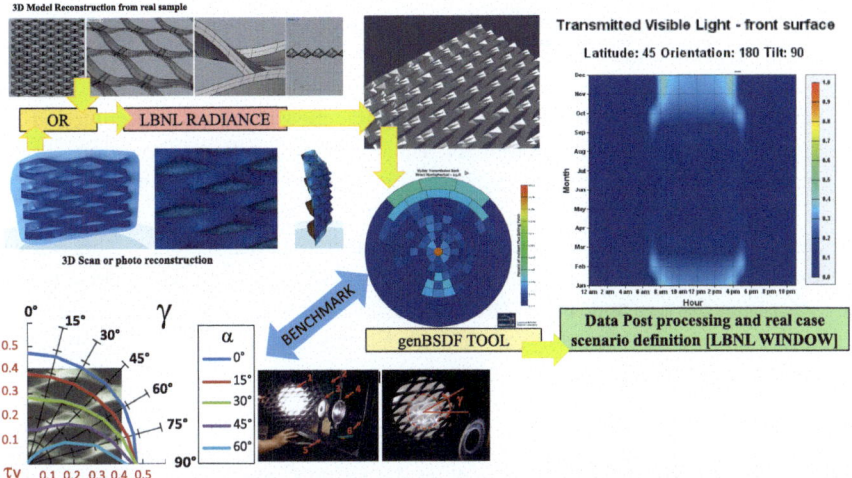

Fig. 3.3 Workflow for the assessment of the angular solar and visual transmittance of a generic shading element sample

detailed and direct investigation. In these cases, two complementary approaches can be employed:

- Measurement of angular optical performance.
- Simulation of solar shading performance.

Having both options available is the best solution, and in such cases, the workflow presented in (Fig. 3.3) can be referred to and can be summarized as follows:

1. *Spectrophotometric Measurements*: Measure the optical properties of the material(s) that constitute the sample to be used.
2. *Geometry Reconstruction*: This can be achieved through computational and parametric geometry generation software or direct reconstruction in 3D CAD by measuring and reconstructing the individual elements of the future shading system to be tested. For particularly complex geometries, this phase can be accomplished through digital scanning. Given the specific nature of this process concerning the object of analysis, it will be detailed in the following subchapter (Sect. 3.2.3).
3. *Import the Geometric Model*: Import the model into physically-based software to create a digital model that considers not only the geometry but also the type of material and its interaction with the incident light/solar flux, including reflectance, transmittance, specularity, and roughness. In this case, the LBNL Radiance application is used.
4. *Processing with Radiance GENBSDF*: Use the GENBSDF tool [17] in LBNL Radiance [18] to create a BSDF (Bidirectional Scattering Distribution Function) file that describes the bidirectional optical behaviour of the model, considering both the angle of incidence and the angle of reflection/transmission. This allows

for handling materials with complex optical properties that cannot be described simply by reflection and transmission coefficients.

5. *Benchmarking*: Compare the angular transmission and reflectance coefficients obtained from the spectrophotometric angular measurements using integrating sphere devices or a goniophotometer. The latter has significant limitations on the maximum texture size of the sample and is mainly suitable for small samples.

6. *Creation of Complex Glazing Systems*: In LBNL Window [19], couple one or more previously created shading layers with transparent systems. This allows exporting the new complex component and tracking its hourly, daily, and seasonal solar transmission and solar factor performance, as well as predicting integration into complex models analysed through Energy Plus [20] and other dynamic simulation tools.

3.2.3   A Speditive Methodology to Acquire Complex Geometries: A Focus on 3D Scanning

In computational design, obtaining digital models for building surfaces is crucial. Whether for architectural visualization or scientific analysis, accurately representing complex 3D forms, textures, and patterns is essential. Surfaces designed for shading purposes demand precise geometrical representation, but the topological definition shifts when considering simulation models. Here, the focus lies not solely on intricate details but on capturing essential characteristics while simplifying overcomplicated surfaces for energy and daylight assessment purposes. This pursuit aims to strike a balance between geometric accuracy and computational efficiency, ensuring that the digital model effectively serves its intended purpose in building simulation.

Digital representation tools empower designers to reconstruct geometries efficiently using a diverse array of digital tools, whether parametric or not, facilitating precise and flexible design exploration. The design methodology enhances precision and efficiency in geometry reconstruction, although could present limitations such as a steep learning curve or software compatibility issues. Digital representation tools play a crucial role following the scanning phases, where refinement of the geometric model is necessary to adapt it to subsequent stages of analysis processes.

3D geometry reconstruction through scanning holds the potential to expedite the transfer of geometries across diverse industry sectors into building technology, particularly within the realm of shading devices. By swiftly capturing and digitizing physical geometries from various fields such as aerospace, automotive, or industrial design, this technique enables rapid prototyping and testing within the context of building technology. This facilitates the adaptation and integration of innovative designs and functionalities into shading devices, streamlining the development process and fostering cross-industry innovation.

Photogrammetry leverages the principles of triangulation to reconstruct 3D geometry from 2D images. By capturing multiple photographs of an object from varying angles, specialized software can discern common points and reconstruct the subject's

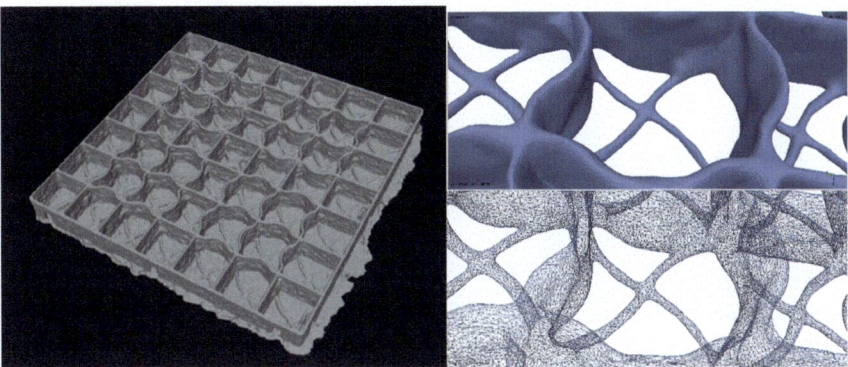

Fig. 3.4 (Left) Photogrammetry reconstruction of a plastic mesh with filled holes; (Right) NeRF refinement and clear void reconstruction

texture and form. This method excels in creating highly accurate and detailed models yet encounters hurdles with objects lacking distinct features or exhibiting regular geometry.

Adequate lighting and precise capture techniques are essential for capturing accurate data. Additionally, comprehensive coverage from various angles enhances the fidelity of the model, ensuring that all aspects of the object are properly represented. Precision in alignment is crucial for subsequent processing steps, facilitating the accurate reconstruction of the scanned object.

However, when scanning complex 3D meshes filled with holes or voids, accurate recognition of the sample is not achieved (Fig. 3.4). To mitigate this issue, random patterns could be introduced behind the sample to assist in alignment and point detection. Despite this enhancement, difficulties persist during the cleaning process, where distinguishing sample points from the background pattern proves to be challenging.

The advent of Neural Radiance Fields (NeRF) significantly enhances 3D scene representation and geometry capture. This innovative approach employs neural networks to interpolate between captured images, synthesizing novel views of 3D scenes and objects. By optimizing network parameters, NeRF predicts both radiance and geometry with an improved level of fidelity.

Following the resolution of the holes issue with photogrammetry, NeRF is utilized to refine the mesh (Fig. 3.4), addressing imperfections and rough surfaces through denoising and polygon count reduction. This process results in a more visually appealing and computationally efficient model.

While the refined model accurately depicts the geometry, it remains ill-suited for simulation in various performance analysis tools. This is primarily due to the high number of points and polygons in the mesh, resulting in computational demands that are often impractical. The linear or quadratic relationship between geometry complexity and computation time exacerbates this issue, rendering the model unsuitable for efficient computational analysis. Thus, despite its fidelity, the model must undergo further simplification or optimization to meet the computational requirements of performance simulation tools.

3.2.4 Static 3d Geometries as Shading Devices: An Example of the Energy Use Assessment

A reference office room is modelled in Energy Plus [20] to estimate and compare the primary energy use for heating, cooling, and electric lighting for south and west orientations. The calculation is carried out for the city of Rome (41° 54′ 00″ N), a Mediterranean city, whose buildings are exposed to severe climatic conditions and high cooling loads in summer. The office unit has a curtain wall façade with a window-to-wall ratio equivalent to 100%. The glazed portion of the façade corresponds to 87% of the window area. The room is equipped with a low-e double glazing unit system (LE), A solar spectral selective glazing unit (SC) is also simulated for comparison purposes. The 3D system is, finally, compared versus a static external venetian blind system; the slat configurations are selected horizontally (VB_0) and tilted 45° (VB_45) against the sun, they also have equal depth and pitch. The thermal and solar properties of the double glazing units (DGU) are computed with ISO 15099 [21] (Table 3.1).

The geometry of the office room is shown (Fig. 3.5). The room has one external wall, it is a single-person office operated from Monday to Friday from 8:00 to 18:00. The internal gains are 10 W/m^2 for electric lighting and 10 W/m^2 for people and appliances. A minimum 500 lx illuminance on the work plane, dimmed and maintained constant in the office unit using a light sensor placed in the middle of the room, The internal air temperature is controlled during working hours between 20 and 26 °C depending on the season. The relative humidity is retained equal to 50% and the airflow is 0.015 m^3/s per person plus a constant infiltration rate of 0.1 Air changes per hour (ACH).

A Single–Zone Variable Air Volume System (VAV) system was considered for the space cooling and heating systems. The primary energy use is calculated from estimated building final energy use, referring to the Italian conversion factors that are 1.00 for gas and 2.18 for electricity. The average nominal coefficient of performance (COP) and nominal efficiency for the gas boiler (η) are assumed respectively 3.00 for Cooling and 0.90 for heating, being typical values for existing buildings.

Concerning the objective of the book, the exemplary case of a metal mesh and metal grid sample is here analyzed as an example of a promising Complex Fenestration System.

Table 3.1 List of the main thermal and optical properties of the glazing system used in the simulations

Glass ID	Uw [W/m^2 K]	g [%]	t$_v$ [%]
LE_2	1.5	49	63
SC_2	1.5	30	69

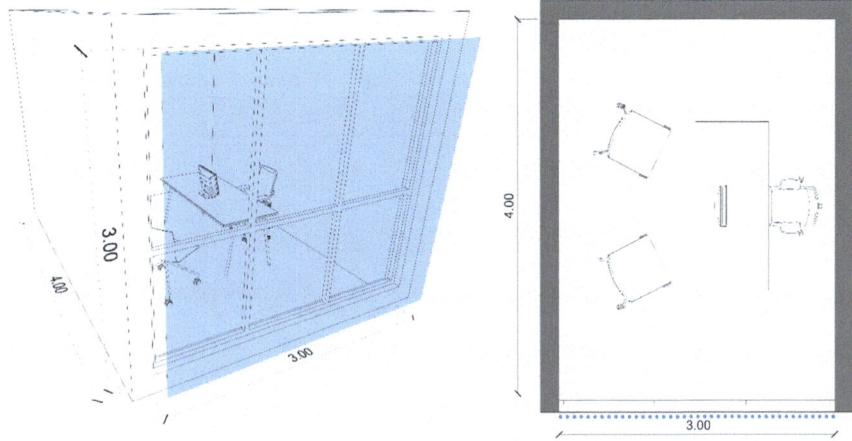

Fig. 3.5 Standard office unit model: External axonometric (left) and plan (right) view and dimension. The highlighted blue panel (left) and dotted line (right) are the shading elements previously presented

The office is shaded with metal mesh and grid systems illustrated in detail in Tables 3.3 and 3.4 as an example of three-dimensional structures. A conventional thermal emissivity ε value of 0.4 and a solar reflectance ρ_e of 0.490 for all the metal surfaces of mesh and grid samples are used.

Mesh samples (Table 3.2) consist only of horizontal wires with a constant spacing for each sample, while Grid samples (Table 3.3) have modular warp and weft spacing.

Mesh and Grid alternatives were characterised in terms of BSDF through genBSDF [17], imported as an XML shading layer and modelled in LBNL WINDOW 7.7 [19] considering the glass system coupled to the shading system as a complex fenestration system (CFS).

The considered alternatives present similar wire dimensions, and among the possibilities, we chose and compared Grid and Mesh alternatives with a comparable openness factor value (OF), to evaluate how this value can be a representative assessment

Table 3.2 Mesh samples with different openness factor alternatives. The wires have a square crossing section

Arrangement	ID	OF [%]	D [mm]	d [mm]	s [mm]
	M1	67	4	2	6
	M2	57	2.6	2	4.6
	M3	50	2	2	4

Table 3.3 Grid samples with different openness factor alternatives. The wires have a circular crossing section

Arrangement	ID	OF [%]	D [mm]	d [mm]	s [mm]
	G1	69	10	2	12
	G2	56	6	2	8
	G3	51	5	2	7

of their shading device performances. Mesh systems (Table 3.3) were considered as a venetian blind alternative with a thickness of every single shading element equal to their depth (d), respecting at the same time the net spacing (D).

The results of the whole simulation set are reported in (Fig. 3.6), reporting the primary energy use for space heating and cooling, electric lighting and the total value, expressed in kilowatt-hours normalized to the usable surface of the office room.

The shading and 3D systems are coupled to LE glazing system, the performance of the latter and of the selective glazing (SC) are also calculated without any additional solar protection systems. It can be observed that the heating use is negligible, the lighting use accounts for about 17% of the total uses and all the rest is ascribed to cooling, which is by far the most predominant energy use.

Despite absolute figures depending on the specific locations, common trends are observed:

- The savings are achieved as a trade-off between consistent cooling savings and an increase in electric lighting use. The relative heating use increases by applying

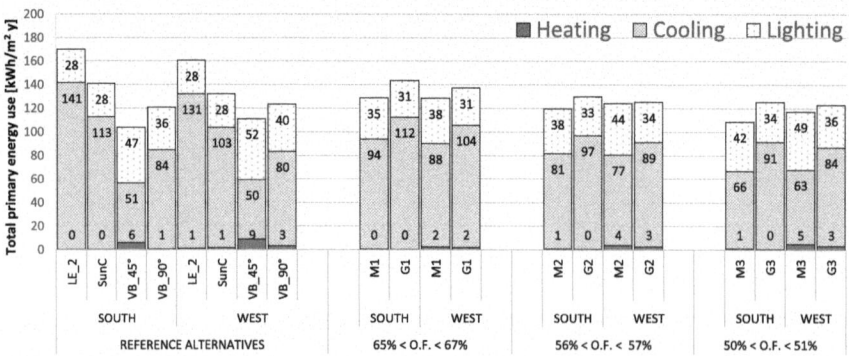

Fig. 3.6 Primary energy use comparison for Heating, Cooling and Lighting for different shading alternatives and orientation for the Rome Scenario

the second skin, however, their contribution to total uses remains marginal, if not negligible, and thus not worth specific analyses.

- The difference depends on the spacing between wires (D value) affecting the wires' mutual self-shading effect.
- 3D systems are more effective in reducing the total energy use when applied on the south façade and with weft elements arranged horizontally. In fact, the system results are more permeable to solar irradiation at the lower incidence angle in the west in summer.
- In all cases but the south-oriented office in Milan, fixed external shading device overperforms 3D systems in terms of total energy uses, this happens despite a relevant increase in electric lighting use.

A short summary of the exemplificative results highlights the effectiveness of shading systems. Total energy use is reduced by 23–39% compared to the LE, depending on the slat tilt and façade orientation. Lower savings are observed when compared to the selective glazing unit, with a peak of 26% for south-oriented 45° tilted slats. The progressive reduction of the 3D systems' openness factor significantly reduces total energy use. This is achieved through a trade-off between consistent cooling savings and increased electric lighting use. For a 50% openness factor (OF), energy savings are 26% for the south and 36% for the west orientation, when compared to LE. These savings drop to 11 and 23%, respectively, when compared to SC.

While applying the second skin increases relative heating use, its contribution to total energy use remains marginal, if not negligible, and thus not worth specific analyses. 3D systems with the lowest OF outperform horizontal Venetian blinds but not the tilted ones. The latter configuration exhibits very similar cooling and lighting energy use, whereas cooling remains the predominant energy use for the horizontal slat and 3D system configurations.

In conclusion, three-dimensional structures such as metal meshes and grids show effectiveness comparable to other static shading devices like brise-soleil. The shading systems analysed numerically have a high openness factor, measured along a plane parallel to the shading surface and per unit area of the shading system. Readers might wonder why lower transparency systems were not used. Their effectiveness depends significantly on geometry and overshadowing dynamics, taking into account factors such as exposure, solar altitude, and incidence angles.

The angular overshadowing effect of the individual elements of the grid and mesh provides an appreciable shading effect from solar radiation, with variable effectiveness depending on latitude, orientation, and time of year. These systems also limit obstruction in the field of view, a common issue with fixed shading systems.

While reducing the openness factor increases the systems' effectiveness in controlling solar gains, a low openness factor can negatively affect user comfort by influencing daylight and outdoor view capabilities. The view through the transparent or semi-transparent surface should be clear, undistorted, and should not significantly alter perceived light conditions. However, these systems alone cannot control other daylight variables, such as glare, thus requiring an additional mobile filtering system, likely on the internal side of the transparent envelope.

It is important to consider that any potential shading performance gap can be addressed by adjusting the glass performance through its solar control and light performances. The aim of this preliminary design investigation is not to suggest that these systems could replace traditional shading systems, but rather to show that they could offer an adequate compromise in providing solar control while supporting the variability of form and material required by contemporary architecture, particularly in complex designs.

3.3   Responsive Building Envelope and the Use of Smart Materials

In contemporary architecture, the concept of a responsive building envelope introduced a paradigm shift. Traditional buildings have always been static entities, with limited ability to adjust their environmental conditions (e.g. through window control). With the advent of new technologies, materials and innovative design approaches, architects are increasingly exploring the potential of responsive building envelopes designed to dynamically interact with their surroundings, adapting to a wider number of environmental factors such as temperature, humidity, sunlight, and wind [22] as will be presented in Chap. 5.

The importance of responsive building envelopes lies in their ability to increase both performance and sustainability by smartly responding to external conditions. These envelopes can optimize energy efficiency, improve indoor comfort levels, and reduce the use of mechanical heating, cooling, and lighting systems. They represent a proactive approach to environmental design, where buildings actively engage with their context to minimize their ecological footprint and enhance occupant well-being [23].

Within this framework materials play a fundamental role in the creation of responsive structures, serving as the building elements that enable dynamic interactions with the environment. Smart materials, in particular, have garnered significant attention for their ability to undergo reversible changes in response to external stimuli [24]. By integrating intelligent materials into architectural designs, architects can endow buildings with a significant level of intelligence and adaptability. These materials enable structures to sense and respond to environmental cues in real time, dynamically adjusting their form, properties, and performance to optimize comfort, efficiency, and sustainability.

Smart Materials feature the capability to change their characteristics if subjected to an external stimulus such as mechanical stresses, temperature/humidity/PH variations and electrical/magnetic field variations. Examples of smart materials are: piezoelectric materials, shape memory alloy (SMA), shape memory Polymers (SMP), magnetoresistive materials, halochromic materials, chromogenic systems (electrochromic materials and thermochromic materials), ferrofluid; photomechanical materials, polycaprolactone (polymorph), self-healing materials, dielectric elastomers (DEs), magnetocaloric materials, thermo bimetals (TBM).

The use of these materials within responsive building envelopes can provide the device itself an adaptive behaviour, depending on the solar radiation intensity, reduces the system complexity and avoids the use of an engine. Direct consequences are both reduced energy requirement and lower risk of failure.

3.3.1 Thermo-Bi-Materials and Shape Memory Materials

Within this wide list of materials, the ones that will be analysed, as currently used for dynamic building envelopes are (Table 3.4):

- Thermo-bi-materials, which can be either thermo-bimetal (TBM) or thermo-moveable-polymers (TMP)
- Shape Memory materials (SMM) specifically Shape Memory Alloy (SMA) and Shape Memory Polymers (SMP).

A Thermostatic Bimetal (TBM) consists of two metal strips bonded together. Since the thermal expansion coefficient of the two materials is different, this component is able to change its curvature with temperature. For this purpose, metals with similar elastic modulus but extremely different thermal expansion coefficients are selected. The "active component" is the metal that, in the couple, has the higher thermal expansion. Generally, this is an alloy containing Nickel, Iron, Manganese, and Chrome in different amounts while the element having minor thermal expansion is called a "passive component" and is usually a Nickel–Iron alloy. The slender the element the higher the curvature but the forces produced are lower and vice-versa [25].

Similar to the TBM is the thermo moveable plastics (TMP) because based on the same physical principle, but as the plastics have higher linear thermal coefficient the movement is higher with lower produced forces [26]. TMPs are cheaper compared to TBM and show a good possibility of being applied as a shading system even if actually there are no shading solutions based on these materials.

Shape memory materials (SMMs) are featured by the ability to recover their original shape from a significant and seemingly plastic deformation when a particular

Table 3.4 Summary of smart materials and their external activating stimulus

Material	External activation stimulus
Thermo Bimetals (TBM)	Temperature, light
Thermo Moveable Plastics (TMP)	Temperature, light
Shape Memory Alloy (SMA)	Temperature, electric current
Shape Memory Polymer (SMP)	Temperature, light, chemical, electric field

stimulus is applied. This is known as the shape memory effect (SME). Smart Materials (SM) can be used in many fields, from aerospace engineering (e.g., in deployable structures and morphing wings) to medical devices (e.g., in stents and filters). The two main classes of materials are:

- Shape memory alloys (SMAs): are the most widespread and studied, their external stimulus is the temperature; however, they are currently quite expensive and their production process implies a considerable amount of waste.
- Shape memory polymers (SMPs): they are less common than alloys, nevertheless they are cheaper (considering the fabrication cost), their external stimulus can be thermal, radiative or chemical and they are generally used in the medical field for small dimensions devices (generally ~ 10–4 m), since they generate small forces.

It is possible to create shape memory composites (SMCs), which include at least one type of SMM (SMA or SMP). The further step for SMM is a newly emerging class of materials, called shape memory hybrids (SMH). They are similar to SMPs, since they are also based on the dual-domain system, in which one is always elastic (the elastic domain), while the other (the transition domain) can change its stiffness remarkably if a right stimulus is presented. Both SMC and SMH with tailored properties/features for a particular application (e.g. self-healing) can be designed. Table 3.5 summarizes the main smart materials that are presented.

Table 3.5 Comparison between TMP, TBM, SMA and SMP

Physical characteristic	TMP	TBM	SMA	SMP
Density [kg/m^3]	900–1400	6000–8000	6000–8000	900–1200
Maximum elongation strain [%]	50–400	–	<10	Up to 800
Working elongation strain [%]	10–200	–	<3	Up to 300
Required stress for deformation [MPa]	1–50	–	50–200	1–3
Stress generated upon recovery [MPa]	1–20	–	150–300	1–3
Deflection coefficient [10^{-6} K^{-1}]	50–200	5–25	–	–
Curvature coefficient [10^{-6} K^{-1}]	10–50	9–45	–	–
Produced stress [MPa]	1–20	2–100	–	–
Transition temperature [°C]	30–90	−20–550	−10–100	−10–100
Recovery speed	1 s—minutes	<1 s	<1 s	1 s—minutes
Processing conditions	<200 °C Low pressure	>1000 °C High pressure	>1000 °C High pressure	<200 °C Low pressure
Cost [€/kg]	5–50	~150	~550	<22

Smart materials (SM), as previously discussed, present significant potential for integration into architectural shading devices as actuators. These materials can enhance the efficiency and reliability of shading systems by minimizing the risk of mechanical failure. However, to fully realize this potential, smart materials must be more than mere substitutes for traditional engines, normally used for the movement of solar shading devices; they necessitate a custom-designed shading model that leverages their unique properties.

An analysis of state-of-the-art materials used in systems incorporating smart materials reveals that the primary types in use are Shape Memory Alloys (SMA), Shape Memory Polymers (SMP), and Thermo-Bimetals (TBM). These materials are employed in various geometries, such as springs or wires for SMAs, custom shapes for SMPs, and foil geometries for TBMs. Table 3.6 provides a comprehensive overview of shading devices that integrate smart materials, categorized by the type of smart material and its geometry.

Table 3.6 Shading systems with SM are divided for material type

Category	References
Shape memory alloy	*SMA Spring based systems*
	1. Self-Adaptive Membrane [27]
	2. ADAPTIVE[SKINS] [28]
	3. SmartScreen (Version C) [29]
	4. Shape Memory Alloy Responsive Façade [30]
	SMA wire based systems
	1. Gill_Project [31]
	2. Shape-changing interfaces [32]
	3. THE AIR FLOW(ER) [33]
	4. Pixel Skin [34]
	5. ADAPTEX [35]
Shape memory polymers	1. Responsive architectural system [36]
	2. Shape Shift [37]
Thermo bi-metals	1. Compliant Shading Enclosure [38]
	2. Glass panel shutter system [39]
	3. Lotus dome [40]
	4. Bloom [41]
Thermo moveable plastics	1. Thermally movable plastic and toys [42]
	2. Thermally movable plastic devices [43]

3.3.2 Integrated Design Process for Dynamic Envelope Components Utilizing Smart Materials

Each area focuses on specific aspects essential to the final design and functionality of a shading device system that takes benefit of the properties of the materials previously listed.

Technology Design

Geometry and movements: The initial phase of the technology design involves integrating Thermo bi-metals (TBM) and Shape Memory Materials (SMM) into a shading device or defining the single materials as a shading material itself. This integration requires careful consideration of the geometry and movements necessary for the device to function effectively and determining the optimal geometry and movement patterns is crucial for achieving the desired shading properties.

Final functional model and shading properties: Based on the possible integration and/or movement analysis, a final functional model is developed. This model defines the shading properties, ensuring it meets the required performance criteria. The focus is on creating a design that maximizes shading efficiency while maintaining aesthetic and functional integrity.

Technical design and components dimensioning: Involves specifying the exact dimensions and technical specifications of each component, ensuring they work harmoniously within the overall design.

Forces displacements: The final step in the Technology Design stream is understanding and calculating the displacements caused by forces within the system. This analysis is essential for determining the mechanical behavior of the materials springs under different conditions.

Working Principles Design

Activation and responsiveness: The focus is on the activation and responsiveness of the shading device. This involves designing the device to respond effectively to environmental stimuli, such as changes in temperature and sunlight, which are the two forces responsible for passively activating the system. The outcome is determining the temperature and activation time required. This data ensures the springs activate at the right moments, providing effective shading when needed.

Daylighting analysis: A critical component of this design phase is the daylighting and solar analysis. This analysis evaluates the impact of the shading device on the natural light entering the space. Ensuring optimal daylighting conditions is essential for enhancing indoor environmental quality and energy efficiency.

Solar and local temperature analysis: The assessment through simulations estimates the thermal behaviour of the shading devices and summarizes the number of occurrences for local activation depending on the overcoming of the threshold for activation due to solar radiation and temperature field.

Dimensioning: The end of the design is the dimensioning of the materials. This step synthesizes inputs from both the Technology Design (forces displacements) and the Working Principles Design (temperature, activation time) to determine the appropriate dimensions for TBM and SMM material.

3.3.3 Concepts for Shading Device Systems Utilizing Smart Materials in Building Façade

Considering the technologies introduced in the previous chapters, it is possible to identify different functional models for applying these various technologies. This chapter does not aim to provide a detailed discussion of every possible application but rather to present some reference scenarios.

The implementation options must account not only for environmental stresses to enable localized and passive activation but also for optimizing the resources involved. Therefore, the two main selection drivers should always be considered:

- Minimizing material usage (kg per m^2 of facade) to avoid significantly impacting costs, particularly for SMM systems, and when these are used directly as shading systems.
- Ensuring that, when smart materials are used as actuators, they achieve maximum action and displacement of the controlled device with minimal deformation, thereby maximizing the shading effect.
- Combining the two aforementioned strategies.

Among the possible alternatives, two project concepts worth investigating:

1. The use of SMA (Shape Memory Alloy) springs as sensor-actuators for a shading system.
2. The use of TMP (Thermo-Moveable Plastic) films as an integrated shading device, utilizing the film's opacity and integrating the material within the glass cavity as a shading element.

In the first case, it is necessary to combine reduced displacement, such as the contraction or expansion of the spring. This movement must then be amplified and directed to achieve the required movement of the shading system. Here, the smart material is confined within an activation black box, a solar absorber that, like a solar collector, receives solar irradiation, causing a local temperature increase. This increase will be proportional to the different activation thresholds of the shape memory alloy material. The spring, connected to a transmission and a roto-translational system, allows the movement defined by the designer. In the specific case shown in Fig. 3.7, the linear displacement of the spring is converted into a rotation that alters the spatial configuration of a discontinuous textile component, transitioning from a double cone configuration (representing maximum system transparency) to a continuous cylindrical shape (representing maximum system opacity) [30, 44]. A manual transmission enabling system override can be easily installed to allow the opaque transition when directly required by the user, depending on particular usage scenarios and applications.

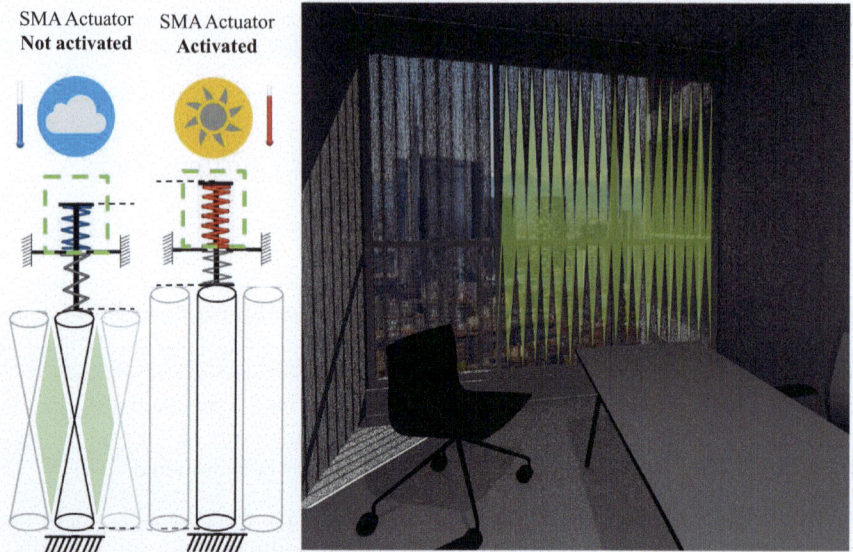

Fig. 3.7 The concept of the SMA spring actuated system. In the green dashed line, the ideal position of the thermal box that drives the activation of the system is depicted, following the conceptual scheme on the left. On the right, an example of activation is shown, demonstrating how local shading and partial activation of the device can be advantageous. The parts highlighted in green indicate the areas where the transition from open to closed configuration is focused

In Fig. 3.8 another dynamic shading mechanism concept is presented. In this latter case, the system is designed to switch between two primary configurations based on temperature changes: an open state when the temperature rises and a closed state when the temperature decreases.

The core component of this system is the Thermo Movable Polymer sheet, which can be rolled and placed either horizontally or vertically. This flexibility allows for the system to be constructed from a single large sheet or multiple smaller sheets. The system is intended to be affordable, customizable, and easy to assemble, while still providing dynamic and effective performance superior to static systems.

When activated, the TMP sheet unwinds to form a noncontinuous, in this case, a screen that blocks sunlight. The closed state, allows light to enter, hence flaps autonomously adjust their shape (openness angle) based on the sun's position and irradiance intensity. The TMP modules, which are black, shift from the a-state rhomboidal shape when non-activated to a d-state square shape (Fig. 3.8) upon activation, effectively doubling their area. This transformation helps modulate light and glare. The index considered in this analysis is the Daylight Glare Probability (DGP) [45]. It indicates the probability that a subject experiences discomfort due to glare, rather than providing a direct measurement or quantification of the phenomenon. A value higher or equal to 45% is considered disturbing.

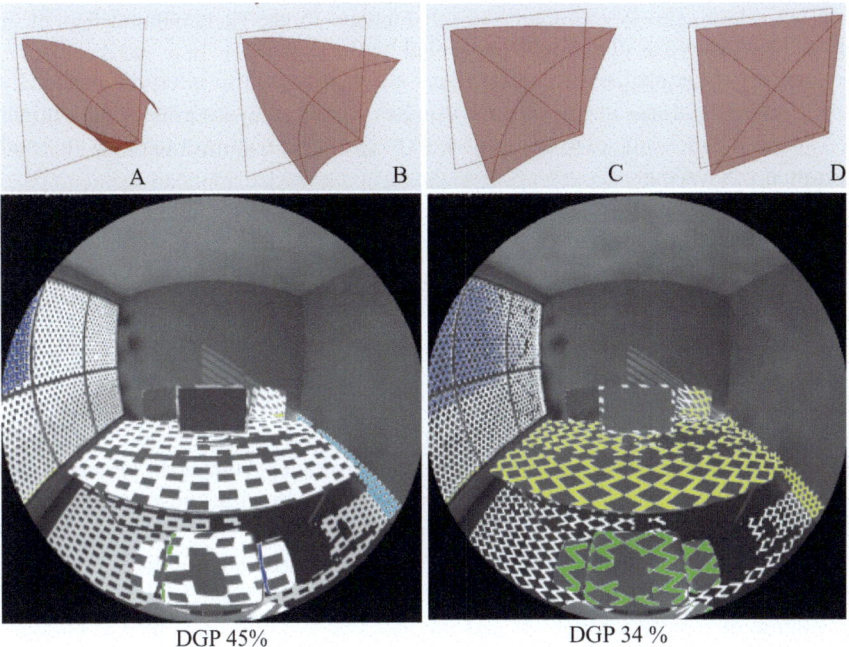

Fig. 3.8 In the upper part, the four transition states of the chosen TMP which is selected as a shading system, while in the lower part, the DGP conditions are depicted with the system deactivated (left) and activated (right)

Enhanced performance could be achieved if the two wings of the system can operate independently, which would require the development of a specialized joint to connect them.

The system can be installed indoors or in the interspace between double-glazed units, though the latter presents challenges related to temperature control and maintenance. External application is not recommended as the TMP sheet is vulnerable to external forces like wind and rain. Manual activation or deactivation is not feasible, and controlling the system can be challenging when needed. Placing the screen in a cavity can enhance performance but also introduces maintenance difficulties and potential operational issues.

Overall, the system is designed to provide a cost-effective, customizable, and dynamic solution for managing daylight and reducing glare. However, it faces challenges in terms of control, durability, and maintenance, particularly in environments exposed to external elements or requiring precise temperature management.

In conclusion, the advantage of these solutions is that each element is formally independent of its adjacent counterparts, allowing for localized activation without the need for complex control strategies, other than linking irradiance to surface/local temperature. This independence also avoids potential obstructions that could interfere with activation. The system can thus self-regulate to ensure maximum benefit or achieve the desired aesthetic effect specified by the designer.

However, intermediate activation states are challenging to interpret, especially during transitional seasons or when the user has specific activation needs. This might necessitate the inclusion of override systems to ensure activation or to bypass normal operation when required.

3.4 Active Systems in Building Envelopes for Enhanced Sustainability

The building envelope plays a pivotal role in the overall energy efficiency and performance of buildings and building's performances can be significantly enhanced through a combination of active and passive strategies even if one of the primary technological challenges in integrating new functionalities into the building envelope is the underdevelopment of certain technologies.

While passive strategies focus on the optimization of the building envelope's thermal properties and the adaptation of the users [46, 47] active ones rely on the integration of new functionalities and services into both new and retrofit building envelopes. Self-sufficient adaptive envelopes (SSAEs) use responsive materials to dynamically modify properties based on environmental stimuli without extra energy input, minimizing complexity [48].

Alternatively, the option should be to include the use of embedded electronic systems, distributed sensors and localized actuators to increase the adaptability and environmental performance of the building envelope [49].

In general, active strategies involve the implementation of advanced technologies to control user behaviour and occupancy, improve the energy use of a building, or improve energy supply, such as integrated renewable energy systems.

Renewable sources can be embedded into the building envelope as Building Integrated Photo Voltaics (BIPV). Studies have demonstrated that BIPV in both opaque and transparent surfaces of the building envelope, can significantly reduce the energy balance of a building [50, 51] and is promoted as a strategy to achieve Zero energy buildings (ZEB) both in the case of new and retrofitted scenarios.

The building envelope can be transformed into a multi-functional element that not only serves its traditional protective role but also contributes to energy generation and environmental sustainability, unlocking a holistic solution to modern energy challenges.

This chapter focuses on the integration of BIPV renewable energy sources with Electric Vehicle (EV) charging stations within the building envelope. The bidirectional EV batteries could be considered as a dynamic storage system representing a forward-thinking approach to modern energy management, while this energy storage capability is crucial for maintaining grid stability and enhancing the building's energy independence.

By addressing these areas, the construction industry can unlock new possibilities for integrating innovative functionalities and services into the building envelope, enhancing both new constructions and retrofit projects.

Factors such as electricity load, location, scale, energy storage costs, and the rise in EV usage all influence the economic feasibility and the Levelized Cost of Energy (LCOE) to ensure economic viability [52]. Unfortunately, the lack of thorough evaluation of the economic and environmental benefits of coupling renewable energy and EV complicates their widespread adoption [53]. Additionally, these benefits vary significantly with different building types and scales, further highlighting the complexity of such integrations, because the economic implications of integrating recharge points into the building envelope are multifaceted.

Dealing with the technology the use of modular construction envelopes and embedded electronic systems could ensure safety, functionality, and adaptability while allowing for efficient upgrades to the building envelope and the usage of adaptable building services. This approach allows the establishment of standards that emphasize the potential of digital technologies like Building Information Modeling (BIM) for a holistic framework covering the building's lifetime from design to operation.

3.4.1 The Concept: A Shared Device for EV Charging Embedded in the Building Envelope

The building sector urgently requires a comprehensive shift towards circular and digital transformation. Product innovation driven by the integration of interconnected buildings and micro-grid neighbourhoods is poised to enhance management and service provision for all users, including vehicle users transitioning into the Internet of Things (IoT) era and smart city environments. This paradigm shift is made possible by newly available, scalable, and modular technologies that cater to user demands such as safety, security, and efficient monitoring and management.

A prototype of the possible integration was then developed in the framework of the project INCASe (Integrated shared ChArge points for Smart buildings) funded by Regione Lombardia for the call of proposals SMART LIVING. The project focused on developing advanced sensor implementations for IoT devices, designed to integrate with building automation systems through modular façade components. This approach primarily targets innovative interventions in existing buildings and the device will interact with plugged-in or registered electric vehicles and smart building automation elements, such as energy storage technologies and personal devices.

The project aimed to unlock dynamic energy performances of the building envelope both to regenerate the existing building envelope and to enhance the newest one, adding new functions and delving into a transformative approach to building technology.

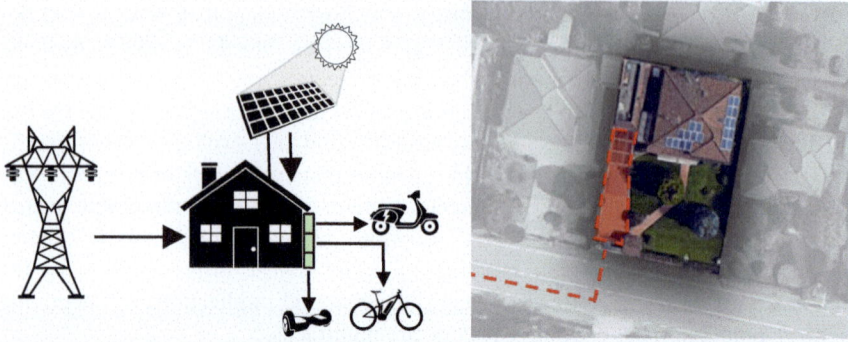

Fig. 3.9 (Left) The general schema of the integration of the device in the building façade and the exploitation of the new functionalities. (Right) The usage of private parking spots as shared space for EV Mobility charging

The primary objective is to seamlessly integrate modular shared charging points for light electric vehicles within façades and in particular modular and prefabricated facades emphasizing an electric power-sharing system that fosters the widespread use of electric vehicles (EVs) (Fig. 3.9). It supports strategic actions to retrofit private buildings into shared spaces for EV mobility, ensuring charging station coverage and reducing costs for public administration (Fig. 3.9). This system offers an alternative to road infrastructure conversion, allowing efficient management of municipal spaces for free or paid parking necessary for recharging. The installation of modular components can facilitate charging availability for various electric mobility systems, including e-bikes, scooters, wheelchairs for disabled individuals, and automobiles. The power-sharing function, based on the status of the electrical system, can enable efficient management of the charging processes, potentially providing tangible economic benefits for the condominium.

The secondary goal is related to systems with local electricity production from renewable sources or BIPV. Control of the energy flow can be implemented to mitigate the mismatch between production and consumption, typically present in both individual and condominium settings, optimizing the rational use of energy. This can form the core of the micro smart grid, serving as the foundational module for the future national smart grid, whose infrastructure is already anticipated at the European level (Fig. 3.10).

The impact of this project is the widespread development of neighbourhood electric vehicles (NEVs) in urban areas and the creation of an interface to a scalable micro-smart grid. This is achieved by providing a condominium recharging system that leverages installed renewable sources.

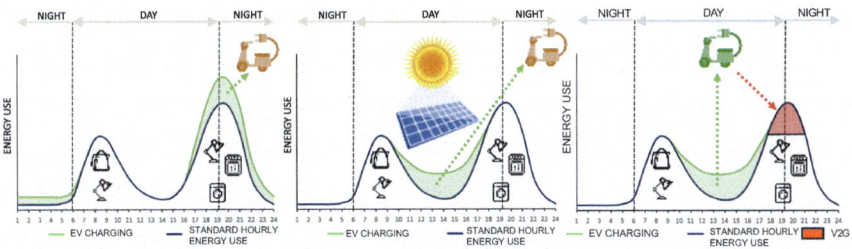

Fig. 3.10 Standard scenario for electric vehicle charging (left); Optimized scenario 1 Taking benefit of the local energy production for EV charging (centre); Optimized scenario 2: Energy consumption delay using the EV as portable storage with V2G performances

3.4.2 The Device

Embedding new functionalities into the building envelope, such as an integrated modular charging point for EVs is a multifaceted process that requires careful consideration of technological, architectural, and practical aspects, ensuring the building envelope while maintaining safety, aesthetics, and structural integrity.

The integration process involves several critical steps, from technological analysis to practical implementation, each of them vital in achieving a seamless and effective integration.

The initial step in this process involves a thorough technological analysis, beginning with the identification of perimeter wall types suitable for system integration. These walls are assessed based on various criteria, including mechanical resistance, thermal resistance, acoustic insulation, fire resistance, system integration capability, replaceability, architectural integration, fire safety, modularity for additional functionalities, and vandalism resistance. Evaluating these factors allows for the identification of the advantages and disadvantages of implementing new functionalities in different wall types, ensuring that the chosen walls can support the new systems without compromising structural or other interrelated performances.

Once suitable wall types are identified the depth of the system that should be installed must align with the standard thickness of facades to ensure compatibility. Then the materials chosen for cladding should be resilient against weather conditions and human-induced impacts, offering mechanical strength and appropriate fire reaction properties. The design should also consider retrofitting compatibility, crucial for integrating new functionalities into existing facades, which is a common scenario in urban areas. (Figs. 3.11 and 3.12).

A detailed analysis of architectural and technological variables needs to be conducted to ensure the system's effectiveness. The depth of the system is adjusted according to the chosen configuration and standard facade thicknesses. Adequate insulation thickness is maintained behind the module to prevent thermal bridging and ensure thermal continuity. The selection of cladding materials involves considering their resistance to atmospheric and human impacts, mechanical strength, and

Fig. 3.11 Construction phases and installation of the devices in an existing building, taking benefit of the existing supporting structure of the ventilated façade

Fig. 3.12 Construction phases and installation of the devices in a new building and a new façade with a dedicated supporting structure

fire reaction properties. Additionally, the surface finish is optimized for colour and reflectance to mitigate the effects of solar radiation on internal temperatures.

The thickness of the cladding is determined based on installation needs, such as slots for power systems and outlets, communication screens, and RFID communication plates. The size of the cladding sheets is also made compatible with standard ventilated facade modules and dimensions, ensuring a cohesive and aesthetically pleasing integration.

The prototype is then mechanically fixed to the structural part of the facade, ensuring a secure and stable integration. Necessary connections are established, integrating the new functionalities seamlessly with existing building systems. Special attention is given to maintaining precise tolerances to ensure correct fitting and uniform gaps relative to the continuous parts of the facade, which is essential for achieving professional and functional integration.

In this specific case, the prototype unlocks the integration into a modular building element (Figs. 3.11 and 3.12) of IoT hardware, enabling the building to interface and communicate with the electrical grid. This transforms the building into a hub for the micro-grid, and serves as a shared charging point for electric vehicles.

The proposed solution is designed for intelligent electricity use and the enhancement of renewable resources within the building, aiming to reduce reliance on the electricity grid for EV charging. Within the context of nearly zero energy buildings (NZEB), the device can:

- Connect to both the electricity grid and private renewable energy sources produced in the building.
- Recharge and manage electric vehicles such as bicycles, scooters, and mobility equipment for the disabled. The number of vehicles charged simultaneously will depend on network availability but will be optimized by power-sharing technologies.
- Connect via open-communication protocols (e.g., Open Charge Point Protocol (OCPP)) for additional app-based services.
- Additional services to improve building operation performance are also anticipated, including security control of the area, remote authorization for building access, and the storage of orders placed with couriers.
- Consequently, the device can interact with both plugged-in and registered electric vehicles, as well as with smart building automation elements, such as energy storage technologies and personal devices.

In conclusion, this chapter sets out to delve into the innovative convergence of sustainable urban mobility and building envelope technology through the lens of the INCASe project. Synergies between urban infrastructure, renewable energy utilization, and digital connectivity are evident as charging points seamlessly integrate with modular building elements, effectively transforming buildings into hubs of sustainable urban infrastructure. This integration enhances energy efficiency in nearly zero-energy buildings (NZEBs) through scalability, adaptability, and compatibility.

- Integration of shared charging points within building envelopes for sustainable urban mobility;
- Technological advancements with hardware and software IoT solutions;
- Impact on urban infrastructure by reducing strain on public charging infrastructure;
- Scalability and adaptability for deployment in various urban settings;
- Broader implications for future urban planning and development;
- New building modules and the envelope as a service.

References

1. Tzempelikos A (2008) A review of optical properties of shading devices. Adv Build Energy Res 2:211–239. https://doi.org/10.3763/aber.2008.0207
2. Fernandes LL, Lee ES, McNeil A et al (2015) Angular selective window systems: assessment of technical potential for energy savings. Energy Build 90:188–206. https://doi.org/10.1016/j.enbuild.2014.10.010
3. Tzempelikos A, Athienitis AK (2007) The impact of shading design and control on building cooling and lighting demand. Sol Energy 81:369–382. https://doi.org/10.1016/j.solener.2006.06.015
4. Heschong L (2002) Daylighting and human performance. ASHRAE J 44:65–67
5. Yu F, Wennersten R, Leng J (2020) A state-of-art review on concepts, criteria, methods and factors for reaching 'thermal-daylighting balance.' Build Environ 186:107330. https://doi.org/10.1016/j.buildenv.2020.107330
6. Al-Masrani SM, Al-Obaidi KM, Zalin NA, Aida Isma MI (2018) Design optimisation of solar shading systems for tropical office buildings: challenges and future trends. Sol Energy 170:849–872. https://doi.org/10.1016/j.solener.2018.04.047
7. Reinhart CF, Voss K (2003) Monitoring manual control of electric lighting and blinds. Light Res Technol 35:243–258. https://doi.org/10.1191/1365782803li064oa
8. Gourlis G, Tahmasebi F, Mahdavi A (2016) Performance simulation of external metal mesh screen devices: a case study. Appl Mech Mater 861:151–159. https://doi.org/10.4028/www.scientific.net/AMM.861.151
9. Palmero-Marrero AI, Oliveira AC (2010) Effect of louver shading devices on building energy requirements. Appl Energy 87:2040–2049. https://doi.org/10.1016/j.apenergy.2009.11.020
10. Chi DA, Moreno D, Navarro J (2017) Design optimisation of perforated solar façades in order to balance daylighting with thermal performance. Build Environ 125:383–400. https://doi.org/10.1016/j.buildenv.2017.09.007
11. Mainini AG, Poli T, Zinzi M, Speroni A (2014) Spectral light transmission measure of metal screens for glass façades and assessment of their shading potential. Energy Procedia 48:1292–1301. https://doi.org/10.1016/j.egypro.2014.02.146
12. Blanco JM, Buruaga A, Rojí E et al (2016) Energy assessment and optimization of perforated metal sheet double skin façades through Design Builder; A case study in Spain. Energy Build 111:326–336. https://doi.org/10.1016/j.enbuild.2015.11.053
13. Blanco JM, Arriaga P, Rojí E, Cuadrado J (2014) Investigating the thermal behavior of double-skin perforated sheet façades: Part A: model characterization and validation procedure. Build Environ 82:50–62. https://doi.org/10.1016/j.buildenv.2014.08.007
14. Alsharif R, Arashpour M, Golafshani EM, et al (2023) Multi-objective optimization of shading devices using ensemble machine learning and orthogonal design of experiments. Energy Build
15. Do CT, Chan Y-C (2021) Daylighting performance analysis of a facade combining daylight-redirecting window film and automated roller shade. Build Environ 191:107596. https://doi.org/10.1016/j.buildenv.2021.107596
16. Konis K, Gamas A, Kensek K (2016) Passive performance and building form: an optimization framework for early-stage design support. Sol Energy 125:161–179. https://doi.org/10.1016/j.solener.2015.12.020
17. Lab LBN (2021) GENBSDF Application of LBNL radiance—computer program
18. Lawrence Berkeley National Lab (2021) LBNL RADIANCE 5.5—computer program
19. Lawrence Berkeley National Lab (2019) LBNL WINDOW 7.7—computer program
20. Energy USD of (2024) EnergyPlus Version 23.1
21. International Organization for Standardization (2003) ISO 15099:2003—thermal performance of windows, doors and shading devices: detailed calculations
22. Luther MB, Altomonte S (2007) Natural and environmentally responsive building envelopes
23. Meyboom A, Johnson G, Wojtowicz J (2011) Architectronics: towards a responsive environment. Int J Archit Comput 9:77–98. https://doi.org/10.1260/1478-0771.9.1.77

24. Sobczyk M, Wiesenhütter S, Noennig JR, Wallmersperger T (2022) Smart materials in architecture for actuator and sensor applications: a review. J Intell Mater Syst Struct 33:379–399. https://doi.org/10.1177/1045389X211027954
25. Guillemin A, Molteni S (2022) An energy efficient controller for shading devices self-adapting to the user wishes. Energy Build 37:1091–1097
26. Sneddon CJ (1996) Toy jewel ornament with thermally responsive cover
27. Gonzalez N (2015) Self-adaptive membrane. https://www.designboom.com/architecture/self-adaptive-membrane-nohelia-gonzalez-shreyas-more-iaac-11-13-2015/
28. Verma S. (2013) Adaptive skins. https://www.designboom.com/project/adaptive-skins-2/
29. Yeadon P (2014) Homeostatic Façade system. https://materialdistrict.com/article/homeostatic-facade-system/
30. Vercesi L, Speroni A, Mainini AG, Poli T (2020) A novel approach to shape memory alloys applied to passive adaptive shading systems. J Facade Des Eng 8:43–64. https://doi.org/10.7480/jfde.2020.1.4700
31. Sandoval J (2012) Gill_project—performance. https://www.youtube.com/watch?app=desktop&v=-X_KxijpJT0
32. Coelho M, Zigelbaum J (2011) Shape-changing interfaces. Pers Ubiquitous Comput 15:161–173. https://doi.org/10.1007/s00779-010-0311-y
33. Lift Architects (2013) Air-flower. https://www.liftarchitects.com/air-flower
34. Sachin A (2008) Pixel skin—robotic membrane. https://www.youtube.com/watch?v=_YF4woob7jc
35. I-Mesh (2017) Adaptex. https://www.i-mesh.eu/researches/adaptex
36. Baseta RSE TE (2014) Responsive architectural system. https://www.designboom.com/architecture/iaac-translated-geometries-08-19-2014/
37. Kretzer M, Rossi D (2012) ShapeShift. Leonardo 45:480–481. https://doi.org/10.1162/LEON_a_00451
38. Vander Werf BD (2009) Elastic systems for compliant shading enclosures. The University of Arizona College of Architecture and Landscape Architecture
39. Sung D, Ramirez A, Phillips L, Michalski J, Xu W, Kang J, Michelle East EH In vert auto-shading windows. https://www.dosu-arch.com/invert
40. Studio Roosegaarde (2010) Lotus. https://www.studioroosegaarde.net/project/lotus
41. Sung D, Wood D (Project Coordinator), Butterworth K, Chen A, Ganis R, Greene D, Michalski J, Sayo Morinaga ES (2011) Bloom
42. Blonder G (2001) Thermally movable plastic devices and toys
43. Blonder G (2005) Thermally movable plastic devices
44. Mainini AG, Speroni A, Vercesi lorenzo (2017) System for shielding and controlling sun light or the light flow coming from artificial sources, especially for application to buildings
45. Wienold J, Christoffersen J (2006) Evaluation methods and development of a new glare prediction model for daylight environments with the use of CCD cameras. Energy Build 38:743–757. https://doi.org/10.1016/j.enbuild.2006.03.017
46. Loonen RCGM, Trčka M, Cóstola D, Hensen JLM (2013) Climate adaptive building shells: state-of-the-art and future challenges. Renew Sustain Energy Rev 25:483–493. https://doi.org/10.1016/j.rser.2013.04.016
47. Sadineni SB, Madala S, Boehm RF (2011) Passive building energy savings: a review of building envelope components. Renew Sustain Energy Rev 15:3617–3631. https://doi.org/10.1016/j.rser.2011.07.014
48. Jacopo C, Jacopo G (2021) Taxonomical investigation of self-sufficient kinetic building envelopes. J Archit Eng 27:03121001. https://doi.org/10.1061/(ASCE)AE.1943-5568.0000504
49. Zarzycki A (2016) Adaptive designs with distributed intelligent systems building design applications. Complex Simplicity 681–690
50. Ramos A, Romaní J, Salom J (2023) Impact of building integrated photovoltaics on high rise office building in the mediterranean. Energy Rep 10:3197–3210. https://doi.org/10.1016/j.egyr.2023.09.178

51. Bennani O, Bensaadout I, Ouassaid M (2016) Positive energy office building: a case study in Casablanca, Morocco. In: 2016 international renewable and sustainable energy conference (IRSEC). IEEE, pp 775–780
52. Hosseini ZS, Khodaei A, Bahramirad S et al (2020) Levelized cost of energy calculations for microgrid-integrated solar-storage technology. In: 2020 IEEE/PES transmission and distribution conference and exposition (T&D). IEEE, pp 1–5
53. Kęska A, Dziubek M, Michalik D (2024) The economic aspects of vehicle operation in the context of electromobility strategies. Combust Engines 196:146–152. https://doi.org/10.19206/CE-172821
54. Mainini AG, Speroni A, Fiori M, Poli T, Blanco Cadena JD, Pizzi R, De Angelis E, Della Torre S, Cattaneo S, Lenzi C, Zanelli A (2020) Regeneration of the built environment from a circular economy perspective rethinking the building envelope as an intelligent community hub for renewable energy sharing. Springer International Publishing Cham 357–361

Chapter 4
Human-Centric Design: Comfort, Well-Being, and Health Cognitive in Building Envelope Design

Abstract In the context of human-centric design, this chapter explores the principles of responsive and cognitive building envelopes. It further discusses the need for broader sustainability and indoor environmental quality practices in building envelope design, emphasizing the need for a paradigm shift and the better articulation of digital, information and communication technologies in building envelope design and operation. The work examines the differences in occupant perceptions of indoor environmental settings, the relevance of capturing such perceptions mainly from the thermal and visual comfort subdomains, the risks of undermining the building envelope, detailed effects of indoor and outdoor proximity, existing and promising methodologies for evaluating and monitoring individual and grouped tolerance levels, and the challenges that the application of these methodologies pose for different stakeholders (e.g., designers, owners, facility managers). Finally, potential solutions to the identified challenges are introduced, before delving into them in the subsequent chapters of this book.

4.1 Introduction

Building design and energy consumption has always been driven by their occupants will and need of comfort and well-being. A concept that has advanced in history from providing a safe indoor environment that enables occupants' survival into an indoor environment that promotes occupants' productivity and commodities (depending on the building typology). In other words, the building design objective has evolved from providing safety, to deliver comfort, and finally to enhance well-being and health. That is, not only ensuring a state of physical, mental, and emotional ease and contentment (or absence of discomfort, pain, or stress) but also promoting better body functioning.

Such improvements have been enabled by the data gathered from indoor and outdoor environmental measurements/monitoring and occupants reported likeness of measured conditions. This has allowed to establish a statistically optimized indoor environmental setting, able to balance different trade-offs among comfort, well-being, energy use and emissions.

4.1.1 Design Bandwidth Due to Occupants' Diversity

Unfortunately, the complexity of the human nature and the variance of subjective evaluations have deemed insufficient the averaged "one-size fits all" approach applied so far in building design, hence also building envelope design [1, 2] (see also Chap. 1—Sect. 1.3). Worsened by the proven mismatch between what has been regarded as energetically optimal building and building envelope operation and registered occupants' preferred function [3, 4]. Moreover, localized discomfort (mainly in the proximity of the building envelope) certainly triggers occupant responses to building and building envelope conditions or operations (Fig. 4.1—generating up to ±25% energy use change) [4].

In response, policymakers, designers, and scientists have shifted towards an approach that considers a more diverse occupancy and/or perception of the indoor environmental conditions across the space (towards an occupant-centric design). In fact, design guidelines and regulations have recently integrated standardized methods that consider: (i) differences in occupants' metabolism (including age and health frailty—ISO 8996:2021 Annex B and C [5]); (ii) differences on the perception of the thermal environment due to direct sunlight (ASHRAE 55:2013 [6]), (iii) indoor wind gust due to ventilation, natural convection, or infiltrations (ISO 17772-1:2017 Annex H [7]); (iv) high luminance contrast difference in the field of view of occupants (ISO 12464-1:2021, ISO 17037:2018 Annex E, ISO 14501 [8–10]); (v) quality of the view to the outdoors (ISO 17037:2018 [9]); among others.

Nevertheless, standards and regulations have introduced these methodologies as optional, unless the complexity of the project requires them or the designer in charge

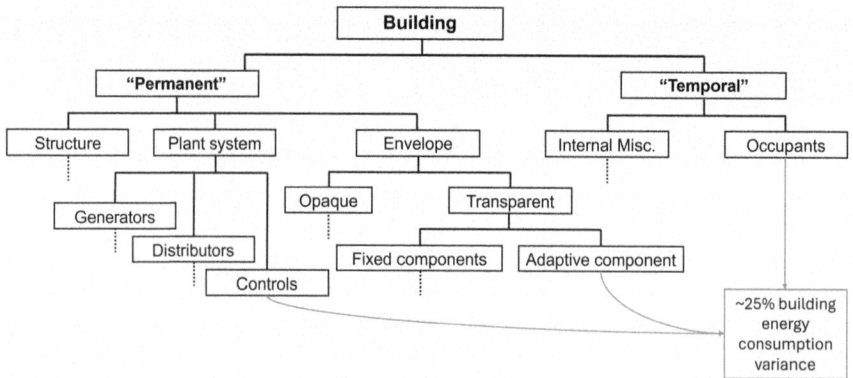

Fig. 4.1 Diagram on occupant-building interaction impact for the European context (extracted and edited from [3])

deems it necessary. Although their complexity and time-burden that impose their application, evidence has been presented on the poor results of post-occupancy evaluations (POE) of the delivered indoor environment (design to be within standard/regulation suggested setting's ranges) [11, 12] and sustained rates of sick building syndrome (SBS) [13, 14].

For instance, Altomonte and Schiavon [12] reported that there is not sufficient statistical evidence from POE results to determine that buildings that follow, or have followed, LEED design guidelines [15] served their occupants better than those who did not. Only slight differences were found better performing in terms of air quality, but worse performance in terms of daylight. In a similar manner, Arens et al. [11] stated that A class (I category) rated buildings under the ISO standards are found to confer no relative satisfaction benefit to individuals or to realistic building occupancies. In addition, there are negligible differences between B and C classes. Also, Boerstra et al. [16] concluded after analysing the Health Optimization Protocol for Energy-efficient Buildings database (HOPE [17] POE in 60 office buildings with >6000 responses) that occupants frequently reported dissatisfaction with the lack or limited options for personal system control to adjust at will their surroundings.

In addition, continuous exposure to indoor environmental settings has been strongly correlated with health affections. To the extent of having, as stated by Rostron [18], the World Health Organization (WHO) [19] recognition of the sick building syndrome (SBS) as "a syndrome of complaints covering nonspecific feelings of malaise, the onset of which is associated with occupancy of certain modern buildings." Such feelings and/or symptoms include sensory (i.e., eyes, nose, and throat) and skin irritation, neurotoxic symptoms, unspecific hyper reactions, odor and taste complaints [20]. And they easily affect, on top of health, building occupants' productivity/learnability and well-being [21].

Hence, motivating a more detailed design and operation approach (e.g., human-centric) that encompasses other domains can potentially capture indoor setting preferences more accurately. In fact, evidence has been published on how further customizing envelope design and operation can result in energy, emissions, and cost savings [4, 22–25]. For instance, extending air temperature HVAC setpoints in mechanically ventilated buildings (assuming a larger heat tolerance), and including personal controls, can potentially achieve energy savings that range between 32%–73% depending on the climate [24]. Up to 56% of energy use in artificial lighting systems can be reduced with automated or frequent light-off switch reminders [4, 25]. And reduced material/system fatigue, maintenance, and replacement by diminishing activation-deactivation cycles of shading systems (−13 to 27%) [23].

4.1.2 Human-Centric Envelope Design, Personalized Monitoring, and Operation

Therefore, it can be meaningful that the building envelope is designed and operated (this includes dynamic external systems to windows) in response to the solicitation of the occupants and the external weather stress. Such dynamism in the responses and performances, coordinated with the building's plant system, can certainly yield proper indoor environmental conditions at lower energy use (e.g., the European Smart Buildings Innovation Community foresees −30% final energy use at building scale [26]). To do so, advancements are in progress upgrading the capacity of designers to estimate more accurate environmental perceptions of building occupants (i.e., more detailed comfort models) and of buildings to acknowledge better indoor environmental conditions and occupants' perception of the experienced space (i.e., cognitive buildings). For instance, the establishment of the Smart Readiness Indicator scheme by the European Commission, and the potential implementation to the future and current building stock [27].

These strategies, coupled with the more detailed design methodologies mentioned in Sect. 4.1.1, could certainly facilitate the integration of occupant physiological and psychological differences (personalized comfort models) into building operational schemes. Thus, fostering a closer relationship between designed, modelled, and real building performance (see Fig. 4.2).

Especially as mostly subjective perception POE has been presented as contrasting with monitored indoor environments resulting in disagreements with established comfortable ranges [29]. In offices in Switzerland, even when indoor environments satisfied established comfortable ranges and thresholds, occupants' satisfaction rates were self-reported always below 50% [29]. Thus, as recommended by Li et al. [30],

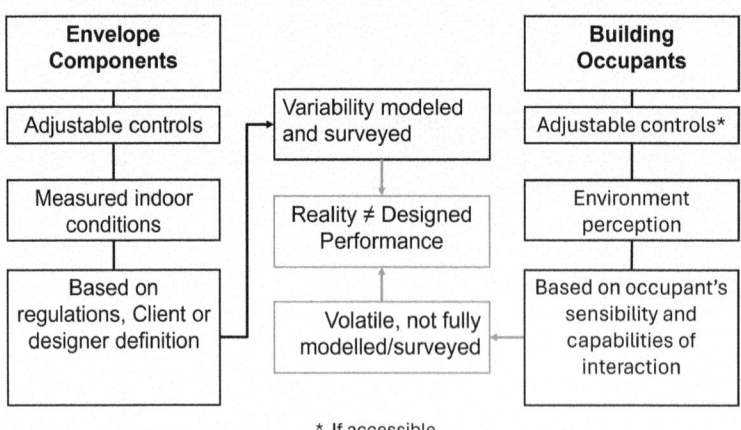

Fig. 4.2 Envelope design and foreseen operation compared to actual activation or controls based on building occupants' perception and sensation [28]

the building acknowledgment of occupant needs should be done continuously instead of periodic one-offs, in a more detailed and standardized manner (allowing to also interpret occupant behaviour patterns), in a human-centric approach, and in a higher interoperability of collected and derived data. That is, increase the sampling rate, collect a larger number of parameters read, capture information of or of the surroundings of the occupant centred, and allow ease of access to be able to adjust building operation or functioning (i.e., building envelope and plant systems).

4.2 Unlocking Building Occupants' Potential for Boosted Envelop and Building Design and Performance

In this context, a paradigm shift is in motion to bridge detail and human-centric building and envelope design to their operation. Such a shift entails moving from studying and monitoring the averaged environmental settings to monitoring the microclimate that surrounds and the humans' responses to such microclimates to estimate perceived environmental quality (IEQ) [31, 32]. This approach allows to implicitly integrate personal preferences, adaptations, and physiological differences [32].

To that end, as mentioned in Sects. 4.1.1 and 4.1.2, detailed envelope modelling can be performed to account for scenarios in which the humans are exposed to particular but rather repetitive scenarios that put them at risk (Sect. 4.2.1). Likewise, human-centric monitoring coupled with envelope and building systems automation can foster sustainable building performance and virtuous occupant behaviours (Sect. 4.2.2).

4.2.1 Specialized Envelope Design for Users

In this context, motivated by reported disturbances (e.g., melted car parts in the City of London in 2013 [33]), existing and surging road safety measures (e.g., BS 5489-1:2020, BS EN 13,201-1:2014, BS EN 13,201-2:2015, BS EN 13,201-3:2015 [34–37]), and elevated thermal perception due to converging solar radiation (e.g., ΔMRT [38, 39]), it has studied how the building envelope shape, materiality, and disposition can potentially affect building outwards and inwards. That is, examining in detail the multi-domain and multi-scale effects and extent that the building envelope shape can have on the proximate outdoor and perimetral indoor areas of the building.

Unfortunately, these issues are often neglected during design (i.e., generalisation while modelling performance). For instance, while designing a building and iterating through envelope design, detailed outdoor thermal comfort and glare risk impacts are not evaluated. Likewise, the effect on, or coming from, surrounding buildings is neglected, or oversimplified (e.g., parallelepiped volumes and averaged or default solar and visual surface reflectance—$\rho = 0.2 \div 0.4$ [9, 40]). Methodologies for

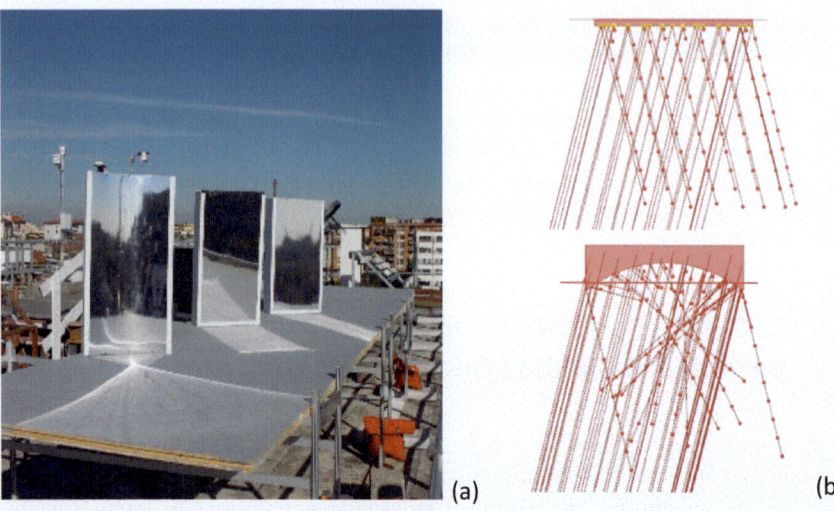

Fig. 4.3 Surveying campaign and ray-tracing comparison of reflected solar radiation from flat and curved facades of tall buildings

evaluating the extent of such effects in similar domains (safety) but for different subdomains are currently enforced in some European countries (e.g., wind comfort and safety in The Netherlands and United Kingdom [41, 42]).

Therefore, an experimental and computer-simulation-based study to deepen the knowledge of the extent of these issues. Concentrating in high-rise buildings with curved facades, for which the effect can potentially be larger [43]. In detail, the work was divided into two: (1) an experimental campaign measuring the magnitude of the sunlight reflected by scale models that reproduce topological archetypes of renowned tall buildings (Fig. 4.3a); and (2) the development and testing of a parametric tool for the preliminary evaluation of the solar reflectance risk potential of a generic complex building shape (Fig. 4.3b).

The experimental campaign consisted of the construction of three 1:100 scaled wooden building prototypes (flat, flat-inclined, and concave), with a south oriented window to wall ratio of approximately 100%, placed on an unobstructed rooftop of a 7-story building (6 stories above ground level), with 15 radially spaced solar irradiance measuring point network (4 rings—2.5, 50, 75 and 100 cm) (Fig. 4.4a). Which allowed to measure the intensity and distribution of the effect of different surface materials (reflective and diffusive) applied to the directly exposed face of the scaled building models. Results suggest that for any building shape with scattering properties, or low reflectance values, solar radiation intensities were found to be below ×1.3 compared to undisturbed horizontal total solar radiation. Meanwhile, for building envelopes with reflective (specular-like) materials, the converged radiation intensities for certain solar positions were measured around ×2 for diffusive surfaces and ×3 times for reflective materials (Fig. 4.4b).

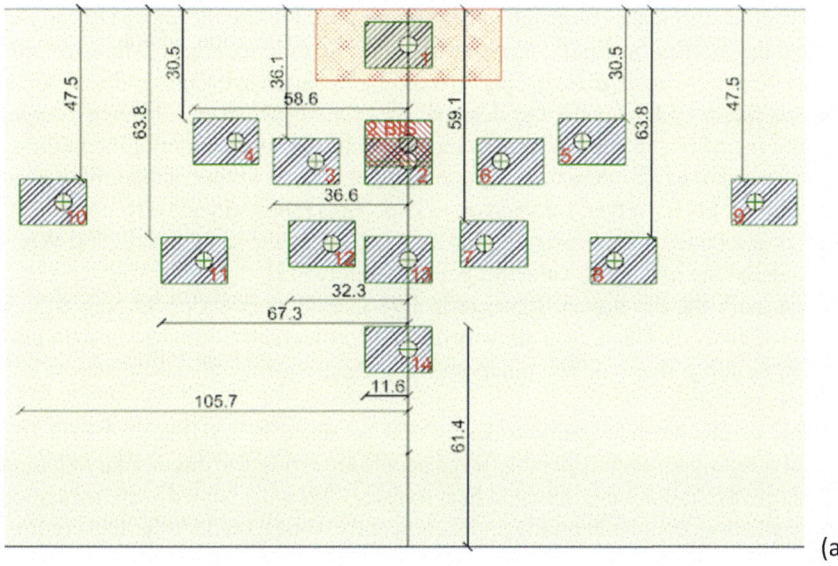

(a)

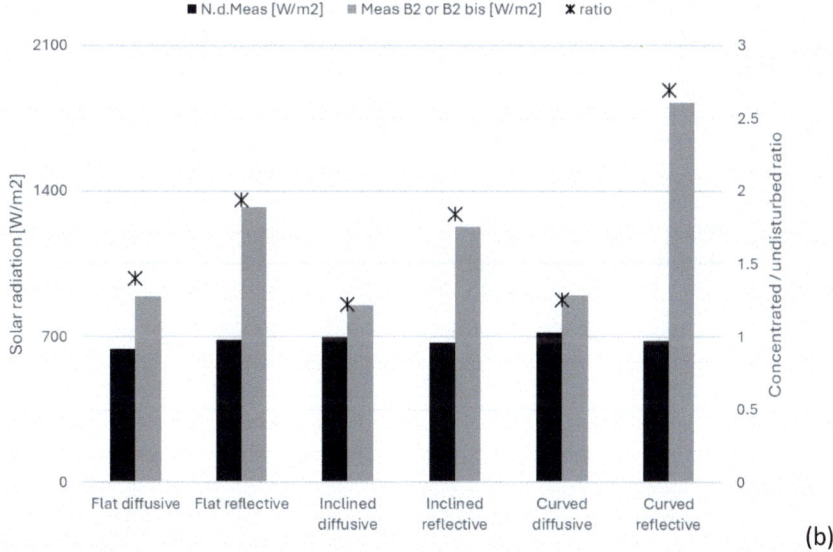

(b)

Fig. 4.4 Details on the **a** surveying campaign layout and **b** the differences between scenarios around noon for 25/09

In specific, during the campaign, on 25/09 at noon there was the highest difference between the undisturbed and concentrated solar radiation measures. Moreover, the specula materials can increase the perceived solar radiation in the proximity of buildings by a factor of 1.5 for flat facades, and of 2 for convex shapes. Having a larger radiation intensity can certainly affect the microclimate just in front of the building while also altering other domains (e.g., driving risks due to veiling or disturbing glare while driving [44]). In fact, for short-term exposures (approximately 10 min), 1500 W/m^2 is associated with strong thermal discomfort. At the same time, 2500 W/m^2 is considered the maximum value for people's safety [45].

Then, with the developed Rhinoceros 3D+ Grasshopper script, leveraged by a parametric analysis (Chap. 5), it allows to identify critical solar altitudes and azimuths for each building shape, to then map annually the frequency of risk scenarios in a selected context.

These results are relevant considering that, as stated in ASHRAE 55:2017 [6], thermal sensation rises considerably when people are exposed to direct solar radiation (i.e., shortwave radiation dominated heat transfer). Which, coupled with intensified solar irradiation, could potentially and significantly worsen outdoor thermal comfort for pedestrians, as well as augmenting surface and air temperatures, influencing urban heat island effects [46]. For instance, for the measured concentrated solar radiation, an additional deltaMRT of 18–20 °C can be obtained, depending on the sky exposure due only to the reflected flux component (utilizing the SolarCal method [6]).

To the knowledge of the authors, such augmented intensity, due to a largely unregulated system radiative property, can worsen indoor thermal conditions and outdoor visual perception. This can potentially affect performance KPIs and operational patterns, and even create new glare risk scenarios for surrounding building users, pedestrians, and drivers.

Hence, SEEDLAB@POLIMI [47] applying human-centric assessment methods (see Sect. 4.1.2), has worked on both indoor and outdoor settings to estimate the effects of direct solar radiation falling on the human body in indoor [38, 48] and outdoor [39] settings. Considering only the indoor settings, and in collaboration with the gfT research group at the University of Cambridge and the Department of Energy at Politecnico di Torino, an analysis and estimation was done on the differences in activation and deactivation patterns of building façade shading systems. Specifically, how a thermal comfort responsive shading system would change its activation pattern considering the thermal perception worsening of direct solar radiation falling on the body of occupants at different locations [48].

In detail, a living lab located in Cambridge (UK) was used as a test case (Luna-Navarro et al., 2018) to estimate the differences in activation patterns. It consists of an office space of 30 m^2, with a south-oriented adaptive glass façade (i.e., electrochromic glass) with a Window-to-Wall Ratio (WWR) of ~0.5. The space is meant to host three occupants, each of them oriented 45° with respect to the south façade, located at different locations (1.20, 2.50, and 4 m from the façade), following the set-up reported by Wienold and Christoffersen [49]. A plan and perspective view of the produced digital model is presented in Fig. 4.5.

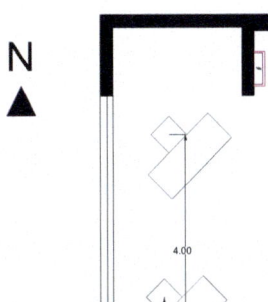 

Fig. 4.5 Test facility schematic plan, and preliminary results on the incident solar radiation in occupants

Different electrochromic glass states were tested on a climate-based analysis (Tsol = 0.02–0.73, Tvis = 0.021–0.66), and for each of them, comfort hours were verified in terms of adjusted operative temperature (Top*—19–26 °C) design criteria [50]. That is, by calculating the longwave and shortwave radiation-driven mean radiant temperature (MRT and MRT* following AHSARE 55 methodology [6, 51]—Eq. 4.1) for each occupant location and then calculating the corresponding operative temperature (Eq. 4.2). This work was carried out with ease, thanks to the parametric workflow laid out by Zani et. al. [38] with Rhinoceros 3D+ Grasshopper + Ladybug Tools. This allows the total short- and long-wave radiation falling on the occupant's body (E_{solar}) to be calculated.

Equation 1

$$MRT^* = MRT_{lw} + \Delta MRT_{sw} \tag{4.1}$$

Equation 2

$$Top^* = \left(hr\,MRT^* + hc\,Ta\right) / \left(hr + hc\right) \tag{4.2}$$

Then, the best thermal comfort condition with each electrochromic glass state was selected (individually and in groups) to generate optimized electrochromic activation levels and patterns. These were then compared to see how different and to what extent there is an increase in activation time and cycles of responsive shading façade systems to maintain indoor thermal comfort. A summary of the differences is presented in Fig. 4.6.

Results showed that traditional optimization methods (CS1) for determining façade system operation are underestimating the effects of direct solar radiation [38, 48]. In general, total activation hours increase significantly (Fig. 4.7), and a detailed

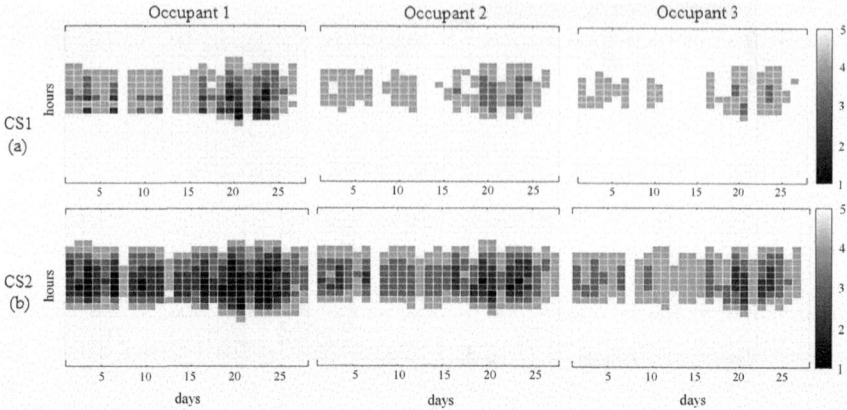

Fig. 4.6 Electrochromic window activation patterns in a summer month based on the evaluated personalized indoor environmental conditions for the hosted occupants. Occupants are numbered based on their proximity to the glazed facade facing south, and activation patterns follow the criteria explained in [48]

analysis should be performed to manage actual perceived and deliverable desirable conditions for most occupants, if not all, of the building occupants (see Fig. 4.7).

These insights will enable designers to anticipate critical scenarios for building users and significantly improve their designs with a more human-centric perspective. By adopting this approach, designers can enhance Indoor Environmental Quality (IEQ) while integrating various domains and subdomains of well-being. Mainini et al. achieved this balance by optimizing energy consumption and daylight provision in retrofitted residential buildings, limiting the thickness of external insulation layers to prevent severe drops in daylight performance [52]. Or, as presented by Imperadori, et. al. analysing different activation cycles and patterns of shading systems following

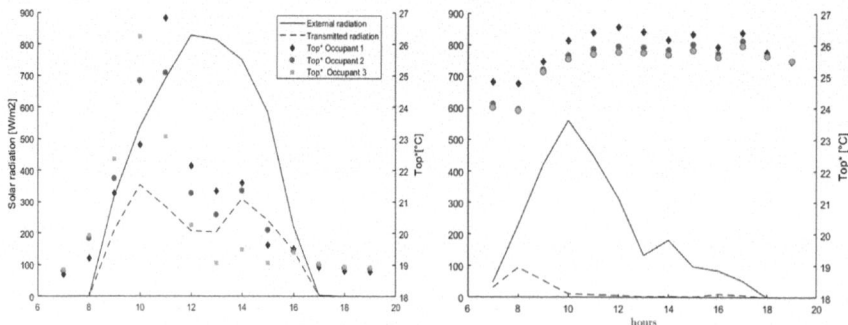

Fig. 4.7 Resulting indoor operative temperatures at each of the occupants' locations, for February (Left) and August (right) considering the effects of direct solar radiation and of the devised electrochromic operation profiles depicted in [48]

different user daylighting preferences (i.e., elders, adults, and youngsters) [23]. Likewise, allowing an informed selection of proper glass panels and interior finishings to boost nonvisual stimulus indoors [53]. Providing results on the melanopic lux intensity (EML) indoor under different combinations of glass panes and interior finishings.

Nevertheless, to guarantee resemblance between designed and operating performance in buildings with this new paradigm (see Sects. 4.1.1 and 4.1.2), such design methodologies must be transferred to building monitoring and actuation systems (e.g., Building Management systems—BMS). As mentioned in Sect. 4.1.2, this can be done by going deeper into monitoring users of the built environment, elaborating that data and capturing useful insights for the building or the façade to respond accordingly.

4.2.2 Customized Envelope Operation from Monitoring Humans

For example, considering that building occupants prefer to sit closer to windows [54–56], within the regionally funded project ELISIR, a cognitive window unit was engineered, built and tested in different building typologies. In specific, Rinaldi et al. and Poli et al. analysed and presented the optimization and utilization of such window unit in an office room. The device presented continuously captures ambiance and behavioural data, to adjust limit thresholds for operation control rules (i.e., window aperture and shading movement) aiming for adaptive and customized comfort system functioning, comprising individual and group dynamics [57, 58].

Differently, to collect data on the micro-environmental conditions surrounding the building occupants, Jin et al. tested an autonomous mobile robot system to collect and evaluate IEQ from spatial–temporal data interpolation [59]. Then, attempting to provide adaptable personalized conditioning systems, Kim et al. proposed new personal comfort models (derived from utilization trends) linked to personal comfort systems (PCS) to directly provide heating and cooling through heat strip fans on chairs [60]. Similarly, Luna-Navarro et al. proposed the use of an IoT toolkit to provide feedback from building users to façade systems (or building management systems (BMS)) by capturing the multi-domain effect of façade operation on IEQ and occupants [61]. Also, Nabilou et al. tested the collection of occupants' feedback through a user-friendly command board for customizing the ventilation efficiency of classrooms for air quality [62]. Equivalently, portable and wearable sensors have been proposed to collect more personalized data and feedback while monitoring their locations and the IEQ in their surroundings to operate façade elements. For instance, Konis and Annavaram presented a mobile (i.e., mobile phone-based) sensing platform pairing physical and subjective data for training and utilizing personalized thermal comfort models to adjust building [63]. And, Allen, et al. designed a wearable pocket badge for capturing illuminance to be used to train and yield responsive operation of façade shading components [64].

In brief, all proven to be driven and enhanced by digital, information, and communication technologies. Finally, recent researchers have opted for directly measuring the human response as a proxy for interpreting IEQ perception, and or the comfort sensation in different domains [3, 31, 32, 65]. Allowing to reduce the influence of subjective domains and subdomains that alter their decision making towards their interaction with the building systems (i.e., the façade). Either with wristband trackers (e.g., smartwatches) for different domains (i.e., thermal, and acoustic) [65], or eye-trackers (i.e., for visual) [3, 28, 66–69].

The above has the potential to enhance building occupants' lives, by continuously adjusting and providing indoor environments that suit their likeness and physiological needs. In fact, making long-term monitoring of both environmental conditions (overall and personal) and physiological responses enables fine-tuning of existing and proposed new comfort models in every domain. Ensuring, however, that such ccc innovations in methodologies and frameworks can be integrated into building design, monitoring, and operation algorithms is not an easy task.

To this end, the experiences gathered while designing and executing surveying campaigns for testing different human-centred strategies in operational buildings have been summarized. In specific, elaborating a literature and experience-based review of human-centred monitoring for understanding daylighting stimulus on building occupants. Measuring in parallel the eye kinematics behaviour and the surrounding environmental state variables in real, immersive and virtual environments. Delivering an easy-to-use diagram for decision making before planning any surveying campaign of visual comfort and perception. The surveying campaigns involved the use of a research-validated eye-tracker (PupilCore [70]), an inhouse built sensing desk for capturing task average illuminance, purposely placed illuminance sensors to monitor illuminance intensity at the virtual display terminal (VDT) and approximately at eye level (see Fig. 4.8), an immersive theatre and a head mounted display (HMD).

Moreover, different daylighting scenarios (e.g., sky conditions, solar altitude and azimuth) and design strategies (e.g., surface properties) were examined to evaluate the capacity of the assessment methods in surveying campaigns to integrate and evaluate the effectiveness or criticality of each design strategy or desired daylighting scenario. Finally, each methodology was scrutinized and compared in terms of time, or effort, for execution, quality, and quantity of data produced to process and validate the output, and the complexity and infrastructure needs for their execution.

These results allow to further improve the thresholds used in operational algorithms of building envelopes, and/or providing further links and more detailed analysis of the actual perception of building occupants. Thus, having a human-centric approach to building envelope operation could yield large energy savings with cost-efficient infrastructure. Coupling data models to occupant actions, motivations and associated domains, can facilitate data analysis and reduce data storage needs and linked costs (see Chap. 6).

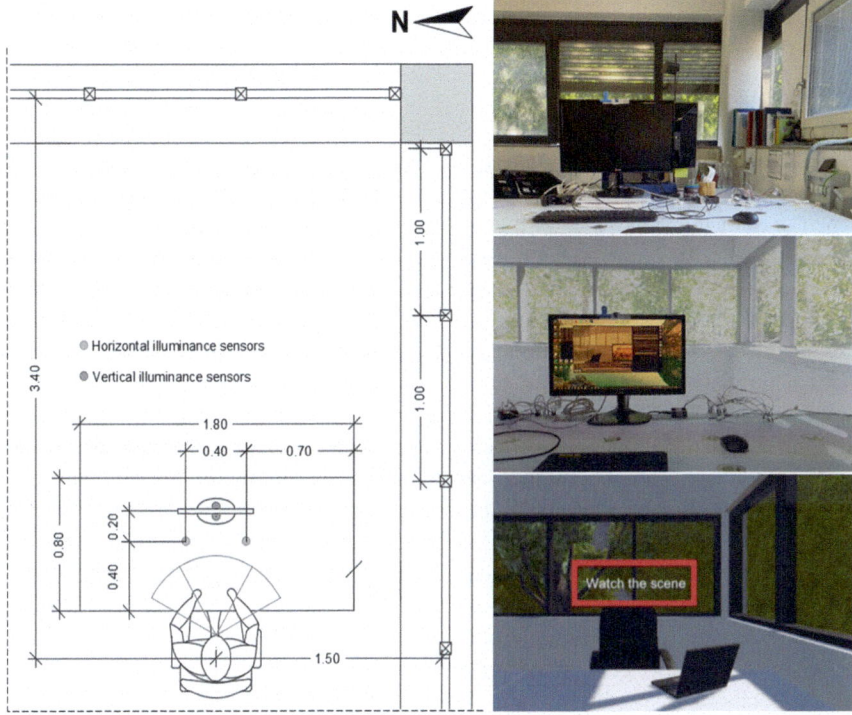

Fig. 4.8 Schematic description and images of an occupant's field of view during the planned surveying campaign, considering environmental physiological responses in real, immersed, and virtual reality scenarios

In this context and concentrating on the scope of this book (Chaps 2 and 6), envelope configuration (i.e., materiality, shape, and morphing) can be better optimized if human-centric performance design principles are integrated into the cost function [26]. To do so, multi-objective optimization techniques can be applied to properly balance involved trade-offs (e.g., energy use vs comfort, individual vs group perception, air quality vs energy use). And such analysis has been eased in the envelope and building design phase with the use of dedicated parametric software and tools (e.g., Rhinoceros 3D+ Grasshopper, Autodesk Revit + Dynamo) that are not only able to test a wide range of possibilities, but also applied numerical optimization methods (see Chap. 5). But also, in the operational phase with rapid decision automation, or suggestions for facility managers, by properly integrating data engineering and machine learning methods to existing infrastructure or with newly installed infrastructure.

4.3   Challenges on Application and Integration for Building Operation

The architecture, engineering, and construction sector (AEC) has been one of the last and slowest in terms of coordinating the integration of digital and communication technologies into operation and performance. That is also related to the low rate of penetration of innovations in built projects [71].

Such apathy to the use of innovation can be related to the high complexity of building projects, as well as the knowledge and skills of practitioners and builders on the application of new technologies [71]. Moreover, they have become a new added construction, and operational, cost and space requirement to consider while managing a building project.

On top of it, to enable or enhance human centric building operation, it is highly complex to have the capacity to read or interpret the perception of building occupants without requesting continuously any feedback from them. Granular time and space data should be collected on the environment and on the users to be able to estimate and forecast potential distress generated by the dynamics of the outdoor, building envelope and indoor environment actors. Sufficient space and time granularity is needed as local discomfort conditions (e.g., radiant temperature gradient, air infiltration, glare) and sudden or short-term comfort disruptions (e.g., reverberation, stagnant air, disability glare, wind gust) are hard to detect with traditional monitoring and data elaboration methods. Furthermore, it can be challenging for a single person to understand and interact with building controls (operable systems and the building envelope) within a room, especially when considering how group dynamics can influence this process [72]. One single disruption can result in virtuous behaviour, or the contrary. For example, a single person or a part of the group suppresses their will to act when they perceive a discomfort condition (see Sect. 4.1.1) or takes the lead towards a virtuous or harmful action for building performance and occupants' wellness.

Wearables, smart elements, virtual sensing, Artificial Intelligence (AI), data modelling and behavioural science surge as promising solutions. However, only a few can overcome the sensing infrastructure redundancy and high cost, together with associated energy use, carbon emissions and cost of data storage [73], that current or future smart buildings are currently facing [74]. That is the case of cost-efficient cognitive or smart elements, virtual sensing and efficient data modelling (Sect. 4.2.2).

Once the cost has been resolved on the sensing side, or if it does not represent a big issue, in the design or running of the building, other issues tend to arise. Data collection processes must comply with the minimum requirements of privacy laid down by the local, regional and national enforced regulations (e.g., General Data Protection Regulation—GDPR). Data protection measures should be applied to all the communication protocols to guarantee cybersecurity, encryption and no breach of personal or sensible information.

New information and communication methods utilized in other sectors come in handy for enabling rapid collection and encryption of data. Computing at the edge allows sensors to preprocess data, allowing less communication between sensors, microcontrollers and servers. It also decentralizes data storage, allows larger control over each data parameter, and facilitates pseudonymization and blockchain integration [75]. Or a simpler alternative is data modelling and structuring, as previously mentioned and further explored in Chap. 6, in which data is only collected for a brief period and continuously overwritten until an action is performed. This allows us to drastically reduce the amount of data stored and facilitate algorithm data processing as less data will be crunched.

The only issue remaining is the flexibility of the building systems and, more in detail, envelope on performing localized corrective actions. For instance, having the capacity of building systems to operate in parallel heating and cooling. Or for the envelope to change its heat, acoustic and visible permeability locally or on certain areas of the building to deliver optimized indoor conditions (see Chap. 5).

References

1. Ahmadi-Karvigh S, Ghahramani A, Becerik-Gerber B, Soibelman L (2017) One size does not fit all: understanding user preferences for building automation systems. Energy Build 145:163–173. https://doi.org/10.1016/j.enbuild.2017.04.015
2. Ghaffarianhoseini A, AlWaer H, Omrany H et al (2018) Sick building syndrome: are we doing enough? Archit Sci Rev 61:99–121. https://doi.org/10.1080/00038628.2018.1461060
3. Blanco Cadena JD, Poli T, Košir M et al (2022) Current trajectories and new challenges for visual comfort assessment in building design and operation: a critical review. Appl Sci 12:3018. https://doi.org/10.3390/app12063018
4. Masoso OT, Grobler LJ (2010) The dark side of occupants' behaviour on building energy use. Energy Build 42:173–177. https://doi.org/10.1016/j.enbuild.2009.08.009
5. (2021) BS EN ISO 8996:2021 ergonomics of the thermal environment—determination of metabolic rate
6. ANSI/ASHRAE (2017) ANSI/ASHRAE Standard 55-2017: thermal environmental conditions for human occupancy
7. ISO (2017) ISO 17772-1 Energy performance of buildings—Indoor environmental quality—parameters for the design and assessment of energy performance in buildings
8. European committee for standardization (2011) BS EN 12464-1—Light and lighting—lighting of work places Part 1: indoor work places 1–57
9. (BSI) BSI (2018) BS EN 17037 Daylight in buildings. Eur Stand
10. (BSI) BSI (2021) BS EN 14501:2021 Blinds and shutters. Thermal and visual comfort. Performance characteristics and classification
11. Arens E, Humphreys MA, de Dear R, Zhang H (2010) Are 'class A' temperature requirements realistic or desirable? Build Environ 45:4–10. https://doi.org/10.1016/j.buildenv.2009.03.014
12. Altomonte S, Schiavon S (2013) Occupant satisfaction in LEED and non-LEED certified buildings. Build Environ 68:66–76. https://doi.org/10.1016/j.buildenv.2013.06.008
13. Gawande S, Tiwari RR, Narayanan P, Bhadri A (2020) Indoor air quality and sick building syndrome: are green buildings better than conventional buildings? Indian J Occup Environ Med 24:30–32
14. Licina D, Yildirim S (2021) Occupant satisfaction with indoor environmental quality, sick building syndrome (SBS) symptoms and self-reported productivity before and after relocation into WELL-certified office buildings. Build Environ 204:108183. https://doi.org/10.1016/j.buildenv.2021.108183

15. USGBC (2018) LEED | USGBC. In: Green build. Counc.
16. Boerstra A, Beuker T, Loomans M, Hensen J (2013) Impact of available and perceived control on comfort and health in European offices. Archit Sci Rev 56:30–41. https://doi.org/10.1080/00038628.2012.744298
17. Cox C (2005) Health optimisation protocol for energy-efficient buildings. Pre-normative and socio-economic research to to create healthy and energy-efficient buildings 16
18. Rostron J (2008) Sick building syndrome: a review of causes, consequences and remedies. J Retail Leis Prop 7:291–303. https://doi.org/10.1057/rlp.2008.20
19. WHO World Health Organization. https://www.who.int/
20. Molhave L (1987) The sick buildings-a subpopulation among the problem buildings?
21. Al horr Y, Arif M, Katafygiotou M, et al (2016) Impact of indoor environmental quality on occupant well-being and comfort: a review of the literature. Int J Sustain Built Environ 5:1–11. https://doi.org/10.1016/j.ijsbe.2016.03.006
22. O'Brien W, Tahmasebi F (2023) Occupant-centric simulation-aided building design. Routledge, New York
23. Imperadori M, Poli T, Blanco Cadena JD et al (2020) Comparison of comfort performance criteria and sensing approach in office space: analysis of the impact on shading devices' efficiency. In: Della Torre S, Cattaneo S, Lenzi C, Zanelli A (eds) Regeneration of the built environment from a circular economy perspective. Springer International Publishing, Cham, pp 381–386
24. Hoyt T, Arens E, Zhang H (2014) Extending air temperature setpoints: simulated energy savings and design considerations for new and retrofit buildings. Build Environ 88:89–96. https://doi.org/10.1016/j.buildenv.2014.09.010
25. Rea MS, Dillon RF, Levy AW (1987) The effectiveness of light switch reminders in reducing light usage. Light Res Technol 19:81–85. https://doi.org/10.1177/096032718701900304
26. Commission E, for Energy D-G, Verbeke S et al (2020) Final report on the technical support to the development of a smart readiness indicator for buildings—Final report. Publications Office
27. Canale L, De Monaco M, Di Pietra B et al (2021) Estimating the smart readiness indicator in the italian residential building stock in different scenarios. Energies 14:6442. https://doi.org/10.3390/en14206442
28. Blanco Cadena JD (2020) Human eye kinematics for adaptable visual comfort assessment. Personalized responsive control strategies to integrate building envelope and artificial lighting. Politecnico di Milano
29. Pastore L, Andersen M (2019) Building energy certification versus user satisfaction with the indoor environment: findings from a multi-site post-occupancy evaluation (POE) in Switzerland. Build Environ 150:60–74. https://doi.org/10.1016/j.buildenv.2019.01.001
30. Li P, Froese TM, Brager G (2018) Post-occupancy evaluation: State-of-the-art analysis and state-of-the-practice review. Build Environ 133:187–202. https://doi.org/10.1016/j.buildenv.2018.02.024
31. Jayathissa P, Quintana M, Abdelrahman M, Miller C (2020) Humans-as-a-sensor for buildings—intensive longitudinal indoor comfort models. Buildings 10:174. https://doi.org/10.3390/buildings10100174
32. Altomonte S, Allen J, Bluyssen P et al (2020) Ten questions concerning well-being in the built environment. Build Environ 106949. https://doi.org/10.1016/j.buildenv.2020.106949
33. BBC (2013) "Walkie-Talkie" skyscraper melts Jaguar car parts. https://www.bbc.co.uk/news/uk-england-london-23930675. Accessed 24 Apr 2024
34. (BSI) BSI (2020) BS 5489-1:2020 Design of road lighting
35. (BSI) BSI (2014) BS EN 13201-1:2014 Guidelines on selection of lighting classes
36. (BSI) BSI (2015) BS EN 13201-2:2015 Performance requirements
37. (BSI) BSI (2015) BS EN 13201-3:2015 Calculation of performance
38. Zani A, Mainini AG, Blanco Cadena JD, et al (2018) A new modeling approach for the assessment of the effect of solar radiation on indoor thermal comfort. Build Perform Anal Conf SimBuild

39. Mirzabeigi S, Khalili Nasr B, Mainini AG et al (2021) Tailored WBGT as a heat stress index to assess the direct solar radiation effect on indoor thermal comfort. Energy Build 242:110974. https://doi.org/10.1016/j.enbuild.2021.110974

40. Zhou W, Chen Y, Liu Y et al (2024) A new method for dynamic load simulation of urban building complexes' heating considering building level occupancy based on shoebox algorithm. Build Environ 261:111710. https://doi.org/10.1016/j.buildenv.2024.111710

41. NEN WC (2006) Wind danger in the built environment. Netherl s Stand Institute, Delft

42. City of London (2020) Microclimate guidelines. https://www.cityoflondon.gov.uk/services/planning/planning-application-requirements/microclimate-guidelines. Accessed 20 Jul 2024

43. Speroni A, Mainini AG, Zani A et al (2022) Experimental assessment of the reflection of solar radiation from façades of tall buildings to the pedestrian level. Sustainability 14:5781. https://doi.org/10.3390/su14105781

44. Donato M, Sessa V, Daniels S, et al (2024) Unlocking spaces for everyone, pp 887–924

45. Council C of S, Committee CSP (2012) Sydney development control plan 2012. Sydney

46. Nuruzzaman M (2015) Urban heat island: causes, effects and mitigation measures—a review. Int J Environ Monit Anal 3:67. https://doi.org/10.11648/j.ijema.20150302.15

47. SEEDLAB@POLIMi Smart envelope energy efficient building and district lab. https://www.seed.polimi.it/. Accessed 27 Jul 2024

48. Luna Navarro A, Blanco Cadena JD, Favoino F et al (2019) Occupant-centred control strategies for adaptive facades: preliminary study of the impact of shortwave solar radiation on thermal comfort. In: Building simulation 2019: 16th conference of IBPSA, pp 4910–4917

49. Wienold J, Christoffersen J (2006) Evaluation methods and development of a new glare prediction model for daylight environments with the use of CCD cameras. Energy Build 38:743–757. https://doi.org/10.1016/j.enbuild.2006.03.017

50. ISO (2005) ISO 7730:2005 Ergonomics of the thermal environment—analytical determination and interpretation of thermal comfort using calculation of the PMV and PPD indices and local thermal comfort criteria

51. Arens E, Hoyt T, Zhou X et al (2015) Modeling the comfort effects of short-wave solar radiation indoors. Build Environ 88:3–9. https://doi.org/10.1016/j.buildenv.2014.09.004

52. Mainini AG, Cavaglià M, Blanco Cadena JD et al (2023) Residential energy retrofits: balancing daylight performance and outdoor view. In: In transition: challenges and opportunities for the built heritage-Book of Abstracts. Edicom Edizioni, p 92

53. Potočnik J, Blanco Cadena JD, Košir M, Poli T (2019) Occupant perception of spectral light content variations due to glazing type and internal finish. IOP Conf Ser Earth Environ Sci 296:012033. https://doi.org/10.1088/1755-1315/296/1/012033

54. Keskin Z, Chen Y, Fotios S (2017) Daylight and seating preference in open-plan library spaces. Int J Sustain Light 17:12–20. https://doi.org/10.26607/ijsl.v17i0.12

55. Gou Z, Khoshbakht M, Mahdoudi B (2018) The impact of outdoor views on students' seat preference in learning environments. Buildings 8:96. https://doi.org/10.3390/buildings8080096

56. Izmir Tunahan G, Altamirano H, Unwin Teji J (2021) The role of daylight on user's seat preferences

57. Rinaldi S, Bellagente P, Ciribini ALC et al (2020) A cognitive-driven building renovation for improving energy efficiency: the experience of the ELISIR project. Electronics 9:666. https://doi.org/10.3390/electronics9040666

58. Poli T, Mainini AG, Speroni A et al (2020) The effect of real-time sensing of a window on energy efficiency, comfort, health and user behavior. In: Daniotti B, Gianinetto M, Della Torre S (eds) Digital transformation of the design. Springer International Publishing, Construction and management processes of the built environment, pp 291–296

59. Jin M, Liu S, Schiavon S, Spanos C (2017) Automated mobile sensing: towards high-granularity agile indoor environmental quality monitoring. Build Environ. https://doi.org/10.1016/j.buildenv.2017.11.003

60. Kim J, Zhou Y, Schiavon S et al (2018) Personal comfort models: predicting individuals' thermal preference using occupant heating and cooling behavior and machine learning. Build Environ 129:96–106. https://doi.org/10.1016/j.buildenv.2017.12.011

61. Luna-Navarro A, Fidler P, Law A et al (2021) Building Impulse Toolkit (BIT): A novel IoT system for capturing the influence of façades on occupant perception and occupant-façade interaction. Build Environ 193:107656. https://doi.org/10.1016/j.buildenv.2021.107656

62. Nabilou F, Maresca M, Mainini AG, others (2023) A multi function and cost effective sensor network system development for indoor air quality monitoring applications. Validation procedure and experimental tests in a school classroom. In: In transition: challenges and opportunities for the built heritage-Book of Abstracts. Edicom Edizioni, p 97

63. Konis K, Annavaram M (2017) The Occupant Mobile Gateway: a participatory sensing and machine-learning approach for occupant-aware energy management. Build Environ 118:1–13. https://doi.org/10.1016/j.buildenv.2017.03.025

64. Allen M, Luna-Navarro A, Blanco Cadena JD et al (2019) User-centred control of automated shading for intelligent glass facades. In: Glass structures and engineering journal

65. Jayathissa P, Quintana M, Sood T et al (2019) Is your clock-face cozie? a smartwatch methodology for the in-situ collection of occupant comfort data. J Phys Conf Ser 1343. https://doi.org/10.1088/1742-6596/1343/1/012145

66. Fugate JM, Fry GA (1956) Relation of changes in pupil size to visual discomfort. Illum Eng 51:537

67. Fry GA, King VM (1975) The pupillary response and discomfort glare. J Illum Eng Soc 4:307–324. https://doi.org/10.1080/00994480.1975.10748533

68. Choi J-H, Zhu R (2015) Investigation of the potential use of human eye pupil sizes to estimate visual sensations in the workplace environment. Build Environ 88:73–81. https://doi.org/10.1016/j.buildenv.2014.11.025

69. Hamedani Z, Solgi E, Hine T, Skates H (2020) Revealing the relationships between luminous environment characteristics and physiological, ocular and performance measures: an experimental study. Build Environ 172:106702. https://doi.org/10.1016/j.buildenv.2020.106702

70. Kassner M, Patera W, Bulling A (2014) Pupil: an open source platform for pervasive eye tracking and mobile gaze-based interaction. In: Proceedings of the 2014 ACM international joint conference on pervasive and ubiquitous computing: Adjunct Publication. ACM, New York, NY, USA, pp 1151–1160

71. Deloitte, Autodesk (2023) The state of digital adoption in construction report 2023

72. D'Oca S, Chen CF, Hong T, Belafi Z (2017) Synthesizing building physics with social psychology: An interdisciplinary framework for context and occupant behavior in office buildings. Energy Res Soc Sci 34:240–251. https://doi.org/10.1016/j.erss.2017.08.002

73. Liu Y, Wei X, Xiao J et al (2020) Energy consumption and emission mitigation prediction based on data center traffic and PUE for global data centers. Glob Energy Interconnect 3:272–282. https://doi.org/10.1016/j.gloei.2020.07.008

74. Hayat H, Griffiths T, Brennan D et al (2019) The state-of-the-art of sensors and environmental monitoring technologies in buildings. Sensors 19:3648. https://doi.org/10.3390/s19173648

75. Chaopeng G, Zhengqing L, Jie S (2023) A privacy protection approach in edge-computing based on maximized dnn partition strategy with energy saving. J Cloud Comput 12:29. https://doi.org/10.1186/s13677-023-00404-y

Chapter 5
Parametric Building Envelope Design and Technology Integration

Abstract The exploration of parametric building envelope design is transforming the field of architecture by integrating advanced materials, technologies, and computational methods. Parametric envelopes are dynamic and adaptable, driven by parameters related not only to aesthetic goals but also to energy efficiency, sustainability, occupant comfort, and cultural value. Aesthetically appealing architecture indeed has to face the complexities of construction taking into consideration the costs to be applied. Innovations in manufacturing processes and materials, such as Cement-Textile Composites, offer promising solutions in this complex field. By allowing the examination and optimization of multiple design configurations, parametric analysis can lead to improved energy performance and reduced operational costs. Additionally, integrating user well-being into the design of parametric building envelopes ensures that indoor environments maintain optimal conditions for human comfort, health, and satisfaction, thus creating more sustainable and liveable buildings. Emphasizing a holistic approach, and combining energy efficiency with user well-being, parametric design holds the potential to revolutionize architecture by creating responsive, sustainable, and aesthetically significant buildings.

5.1 Introduction

Building envelopes has always been a field of experimentation for new materials and technologies. In this experimental context, the parametric design has been included. Parametric building envelopes are dynamic and adaptable layers for the outer skin of a building, where the geometry, performance, and behaviour of the envelope are driven by parameters. These parameters can be related to environmental conditions, material properties, structural requirements, or aesthetic goals, and they allow for the creation of complex, efficient, and optimized shapes. Some of the key aspects that characterize them are:

© The Author(s), under exclusive license to Springer Nature Switzerland AG 2024   81
A. G. Mainini et al., *Unlocking the Potential of Building Envelopes*,
PoliMI SpringerBriefs, https://doi.org/10.1007/978-3-031-75298-8_5

- *Dynamic behaviour*: The envelope can exhibit dynamic behaviour typically for aesthetic or performance reasons. In terms of aesthetics, the envelope can change various parameters such as shape, colour, and brightness, and can also alter its degree of transparency/continuity.
- *Optimization and adaptation*: Parametric design tools, often powered by algorithms and computational methods, help in optimizing the building envelope for better performance. The envelope can respond to various parameters such as natural lighting conditions, wind patterns, and thermal performance. This means the building facade can change or be optimized based on real-time data or predefined criteria. This can lead to reduced energy consumption, improved comfort for occupants, and better overall sustainability.
- *Complex Geometries*: Parametric design allows for the creation of complex and innovative geometries that would be difficult or impossible to achieve with traditional design methods. This can result in unique and visually striking building facades.
- *Multi-discipline integration*: By incorporating parameters from different fields such as structural engineering, environmental science, and material science, the building envelope can be designed to meet multiple objectives simultaneously.
- *Customization and Flexibility*: Parametric models allow for easy modifications and customizations. Changes in parameters can quickly lead to new design iterations, making the design process more flexible and responsive to new information or requirements.

Having a parametric building envelope can be highly beneficial both for building occupants and for people outside of it due to several key reasons. These advantages span across energy efficiency, sustainability, comfort, well-being, and aesthetic and cultural value, making parametric design a revolutionary approach in modern architecture.

5.2   Energy Efficiency and Sustainability

One of the most significant benefits of parametric building envelopes is their contribution to energy efficiency and sustainability. By optimizing the building's response to environmental conditions such as sunlight and wind, parametric envelopes can significantly reduce the need for artificial heating, cooling, and lighting. This leads to lower energy consumption and a smaller carbon footprint, which is crucial in the fight against climate change. The ability to dynamically adjust to the environment ensures that the building operates at peak efficiency, using resources only when necessary.

Moreover, parametric design facilitates the use of sustainable and locally sourced materials. By integrating these materials into the design process, architects can promote environmental stewardship and reduce transportation emissions. The emphasis on sustainability extends beyond the operational phase, considering the entire lifecycle of the building materials, from extraction and production to eventual disposal or recycling. This holistic approach ensures that the building's environmental impact is minimized from start to finish.

Comfort and Well-Being

The enhancement of comfort and well-being for occupants is another critical advantage of parametric building envelopes. By adjusting to climatic conditions, these envelopes can improve indoor air quality, natural lighting, and thermal comfort. Such improvements have a deep impact well-being of occupants, contributing to a more pleasant and productive indoor environment. Natural lighting, in particular, has been shown to improve mood and reduce the incidence of illnesses associated with poor indoor air quality.

Dynamic features such as adjustable shading devices and natural ventilation systems further enhance indoor comfort. These systems help maintain comfortable indoor temperatures and reduce reliance on mechanical heating and cooling systems. The adaptability of parametric envelopes ensures that buildings can provide optimal comfort in various weather conditions, contributing to the overall well-being of their inhabitants.

Aesthetic and Cultural Value

Parametric design also brings significant aesthetic and cultural value to buildings. The ability to create unique and aesthetically pleasing building facades can transform structures into landmarks, fostering a sense of pride and identity within the community. Architectural innovation through parametric design enables the creation of forms and patterns that were previously unimaginable, pushing the boundaries of what is possible in building design.

Furthermore, parametric design allows for buildings to harmonize with their surroundings. This contextual integration ensures that new structures respect local architectural styles and cultural heritage while incorporating modern elements. The blend of tradition and innovation can enhance the cultural fabric of a community, creating spaces that are both functional and meaningful. In doing so, parametric building envelopes not only improve the visual landscape but also strengthen the cultural connections within a community.

Thus, the benefits of parametric building envelopes extend far beyond mere aesthetics. By enhancing energy efficiency, sustainability, comfort, well-being, and cultural value, parametric design offers a comprehensive approach to creating better buildings for people and communities. This innovative design methodology holds the potential to revolutionize the built environment, making it more responsive, efficient, and connected to the people it serves.

Economic Benefits

The economic implications of parametric building envelopes are substantial. While the initial costs associated with their design and implementation might be higher, the long-term savings achieved through reduced energy consumption and higher user comfort can be significant. These savings make the investment worthwhile, particularly when considering the lifecycle of the building. Additionally, buildings equipped with advanced, efficient, and aesthetically appealing envelopes often see an increase in property value. They attract premium rents and higher sales prices, enhancing the financial returns for property owners and developers.

Resilience and Adaptability

Parametric envelopes excel in their resilience and adaptability, crucial in today's rapidly changing climate. Designed to withstand extreme weather events, these envelopes improve the durability and resilience of buildings, safeguarding communities against the impacts of climate change. Furthermore, parametric designs offer unparalleled flexibility for future modifications. As community needs evolve, buildings with parametric envelopes can be more easily adjusted or expanded, unlike traditional static designs that might require extensive renovations or are unable to adapt smoothly.

Community Engagement and Participation

Community engagement is another profound benefit of parametric design. The design process can actively involve community input, resulting in buildings that more accurately reflect the needs and preferences of residents. This inclusive approach not only fosters a sense of ownership and pride among community members but also enhances the functionality and relevance of the built environment. Additionally, the innovative nature of parametric design provides educational opportunities, inspiring communities and sparking interest in sustainable architecture and engineering. It acts as a catalyst for educating the public about the benefits of innovative, sustainable design practices.

Health and Environmental Benefits

Health and environmental benefits are also integral to parametric building envelopes. These designs promote natural ventilation, significantly enhancing indoor and outdoor air quality by reducing reliance on mechanical ventilation systems that consume energy and often rely on fossil fuels. Improved air quality has direct positive effects on public health, particularly in urban environments. Moreover, by incorporating elements like green roofs and walls, parametric envelopes help mitigate the urban heat island effect. This not only makes cities cooler and more comfortable but also contributes to broader environmental goals such as reducing carbon emissions and enhancing urban biodiversity.

5.3   Background

With the advent of parametric design, a true revolution has taken place by integrating advanced computational methods to enhance buildings' envelope aesthetic and functional aspects. Parametric design is a computational approach that uses algorithms to define and manipulate multiple parameters, enabling the creation of flexible and precise models. By adjusting input variables, it can explore numerous configurations efficiently. This method enhances creativity, and accuracy, and reduces design iteration time, making it essential in modern architecture and engineering [1]. This changed the architectural landscape by enabling more flexible, efficient, and creative design processes. Using algorithms and computational tools, architects and designers can generate complex geometries and explore a multitude of design options quickly [2]. Parametric design indeed allows to customize building envelope shapes, to optimize environmental performance, aesthetic appeal, and material efficiency [3]. Consequently, a multitude of parameters can be detailed managed and optimized such as daylight, airflow, and energy efficiency, leading to more adaptive and occupant-centric designs as previously presented in Chap. 2.

In the era of Industry 4.0, characterized by the fusion of digital, physical, and biological systems, parametric design stands as a crucial enabler of innovation [4]. The integration of the Internet of Things (IoT), artificial intelligence (AI), and big data analytics within the construction industry allows for the creation of smart, responsive building systems. Parametric design supports this by providing the necessary flexibility and precision to adapt to real-time data and changing requirements. It facilitates the design of buildings that are not only more efficient and sustainable but also capable of interacting with their environment and users dynamically. It is a crucial process for building envelope design from the production perspective due to its ability to streamline the design-to-manufacturing process, enhance efficiency, and enable precision in construction.

Khamis et al. [1] explain that the parametric design process begins with the conceptual stage using tools like Grasshopper and Rhino3D, which then transitions into detailed execution drawings via BIM (Building Information Modelling) software like Revit. This workflow facilitates the creation of multiple design alternatives, aiding decision-making and ensuring the selected design is optimized for performance. The use of parametric design also allows for digital fabrication techniques such as 3D printing, which aids in the creation of precise physical models. Rizi et al. [5] highlight how parametric design in building envelopes can optimize daylight and ventilation performance, contributing to energy efficiency and occupant comfort. Manavis et al. [6] demonstrate how parametric tools can generate innovative architectural structures by leveraging computational design techniques, promoting sustainability and urban greening. Winter Han [7] discusses the application of parametric optimization in selecting optimal wall and roof configurations, which significantly reduces energy demand in high-rise residential buildings. Yılmaz and Mendilcioglu [8] emphasize that parametric design facilitates the creation of culturally and environmentally adaptive structures, enhancing both functionality and aesthetic appeal.

Parametric design enables unprecedented levels of customization and optimization, which can lead to significant improvements in building performance and resource efficiency. Furthermore, parametric design allows for greater collaboration and integration across disciplines, as data-driven models can be easily shared and iterated upon. As technology continues to advance, the integration of AI and machine learning with parametric design promises to unlock even more innovative solutions and applications, driving the future of architecture and construction towards greater adaptability and sustainability. This continuous innovation helps architects and designers to push the boundaries of what's possible, ensuring that the built environment remains responsive to the evolving needs of society.

Despite its advantages, parametric design is not without limitations. The main problems of parametric design for building envelopes include complexity in optimization, integration challenges, the need for advanced technical skills, and significant implications for manufacturing processes. Brakke and Velasco [9] argue that parametricism can be limited to style and aesthetic issues, requiring iterative processes for effective problem-solving in contemporary design. Hou et al. [10] highlight difficulties in constructing optimization models, lack of dynamic visualization, and integrating these models into the actual design process. Lin et al. [11] point out that conventional design methods struggle to adapt to dynamic climate changes and advanced performance conditions, necessitating an integrated parametric approach. Steinø [12] discusses barriers such as lack of knowledge among practitioners, time constraints, and inadequate understanding of practice workflows.

5.4 Parametric Building Envelope Shape: From Customized Design to Optimized Production

In recent years, building envelope design has progressed incorporating complex geometries and surfaces that are both visually captivating and tailored for specific user needs. This evolution considers exterior architectural surfaces, interior surfaces, and even individual striking components embedded within standard surfaces. The transition toward bespoke optimization approaches has necessitated a separation from standardized systems and components, favouring instead single-use, customized, and optimized solutions. This concept is further supported by the European community through the "New European Bauhaus," which envisions and collaboratively constructs a sustainable and inclusive future without neglecting aesthetic considerations [13].

As a direct outcome of this paradigm, there is an increasing shift away from standardization, leading to the production of custom or bespoke architectural surfaces in limited quantities. This transition, while presenting significant challenges for designers and manufacturers, meets the rising demand for complex designs. Architects and designers now seek unique solutions tailored to individual projects and user requirements [14]. Parametric design, as previously presented, often involves the implementation of complex geometries, including non-Euclidean geometries such as elliptic, hyperbolic, and fractal geometries [15], as well as organic and non-standard configurations [16]. This design approach is facilitated by specific digital technologies that effectively model designers' intentions. Michael J. Ostwald has outlined the historical development of architecture as a parallel evolution between the design process and enabling technologies. Enabling technologies (ETs) comprise any set of tools, techniques, or protocols necessary to support a given design process [17].

Accessibility to suitable technology is critical for assessing the feasibility of a design process, as the limited implementation of specific design features often stems from technological limitations, high production costs, scarcity of machinery, or a shortage of skilled professionals, which hinder the overall feasibility of certain design choices.

Double-curved surfaces, while highly appreciated and successfully implemented in various high-budget designs, notably in building envelope design, are among the design features infrequently addressed in practice due to these technological constraints [18]. The realization of double-curvature designs requires sophisticated machinery and a skilled workforce, necessitating significant financial investment. This makes such designs less viable for smaller-scale projects with limited resources [19]. To address this gap, enabling technologies have been thoroughly researched in recent years, focusing on digital manufacturing technologies.

In building envelope design, the outer appearance of a building is often achieved through a cladding system. Cladding systems involve applying one material over another to provide a skin or layer over the building, defining its outer shape and finishing material [20]. Non-loadbearing claddings can be implemented using outer lightweight panels in conjunction with an inner structural framework. This setup allows for flexibility in defining the shape and material of non-structural panels, with materials ranging from brick, stone, metal (aluminium or steel), timber (wood), glass, concrete, ceramic tiles, vinyl siding, fibre cement, various composite materials (aluminium composite panels, fibre-reinforced plastic) [20, 21] (Table 5.1).

Multiple manufacturing processes have been developed to produce cladding panels with different geometries, both planar and three-dimensional as presented in Table 5.2. Three-dimensional shapes include meshes of planar polygons or curved surfaces with single or double curvature. Building designs that incorporate complex cladding geometries are generally referred to as "free-form" constructions. Free-form construction often pushes the limits of current technological capabilities, resulting in highly complex manufacturing needs. Due to this complexity, the sustainability of free-form construction is frequently questioned [22]. Kavuma et al. noted that free-form projects often run behind schedule and over budget, suggesting that fabricating free-form components significantly impacts timely completion and cost issues, largely dependent on the involved personnel [23]. The complex manufacturing processes required for free-form buildings reduce flexibility in overcoming contingencies such as client decisions, delays, or project changes, leading to cost and time overruns.

The manufacturing process for free-form building envelopes typically focuses on the outer panels through a process known as "panelisation" [24]. Panelization simplifies a free-form geometry into a finite series of sections (panels), as shown in Table 5.1, which can be individually manufactured and assembled on-site. The panel geometry is usually planar, achieved through various techniques such as triangulation, primitive approximation, fitted rotational surfaces, principal curvature meshes, and developable strip models [25], resulting from an overall approximation of the proposed design.

Manufacturing panels with true free-form three-dimensional geometry (e.g., double curvature geometry) is achievable through a limited number of manufacturing

Table 5.1 Possible façade base geometries and their penalisation

Geometry	Key features
	Curvature: No curvature *Panel type: All equal* *Quadrability: maximum*
	Curvature: Single *Panel type: Equal for row* *Quadrability: good*
	Curvature: Double *Panel type: Few types* *Quadrability: Acceptable*
	Curvature: Double *Panel type: All different* *Quadrability: Based on geometry*

technologies, particularly digital manufacturing technologies utilizing CNC production machinery. Materials suitable for panel production include plastic, glass, metal [26], and specialized materials such as fibre-reinforced polymers (FRP) [27]. Metal panels are generally produced through bending, die forming, single-point forming, and similar processes [26]. However, cold bending methods cannot produce precise

Table 5.2 Comparison between standard production processes for complex shapes. * = low, ** = medium, *** = high

Process category	Process	Relevant materials	Cost-effectiveness of the process	Cost-effective initial investments	Raw material costs	Geometry complexity	Surface quality	Production speed
Forming	Thermoforming	Thermoplastic	**	**	**	*	***	**
	Forming with mould	Metal, wood, plastic	*	**	**	**	**	***
	Autoclave moulding	Composite materials	*	*	***	**	***	*
Casting	Mould casting	Fluid materials	**	**	*	**	*	*
	Casting on flexible adjustable mould	Fluid materials	**	*	*	*	*	**
Milling	CNC milling	Virtually all	***	**	**	***	**	**
3D printing	Fused deposition modelling (FDM)	Plastics	***	**	*	***	*	**
	Selective laser sintering (SLS)	Nylon, polyamide	**	*	**	***	**	*

double-curvature geometries above a certain complexity level [28]. Advanced materials like glass and fibre-reinforced materials, such as fibre-reinforced plastic (FRP) and glass-fibre-reinforced concrete (GFRC), require production processes involving moulds [29]. This necessitates a focus on the production, utilization, and waste generated by mould implementation [30]. Free-form moulds are typically produced using digital manufacturing machinery employing both subtractive and additive methods.

However, current manufacturing processes are often inadequate for economically and sustainably producing complex shapes and surfaces in small quantities. In this context, mainly associated with small-scale production, the priority Is typically given to the complexity related to the production and feasibility while the employment of principles based on Design for Disassembly and Deconstruction (DfD) unfortunately becomes secondary. To address these challenges, designers and manufacturers are exploring innovative materials and manufacturing processes that facilitate the production of complex shapes and surfaces in small quantities. As the demand for sustainable building materials and processes grows, in alignment with UN Sustainable Development Goals 9 and 12, there is a need for innovative technologies that enable the production of custom architectural surfaces more sustainably and efficiently.

Based on the described technology, it is therefore possible to address the main issues and limitations associated with parametric building envelopes with complex shapes as follows:

- Complexity in Fabrication: Parametric designs often involve intricate and complex geometries that require advanced fabrication techniques. These methods are costly and demand specialized equipment and expertise, which may not be readily available in all manufacturing settings.
- Material Limitations: The materials used must be compatible with the chosen fabrication techniques. Some complex designs may require materials that are difficult to work with or are not suitable for the intended application due to their mechanical properties, durability, or environmental performance.
- Production Costs: The high level of customization and precision in parametric design often leads to increased production costs. The need for advanced manufacturing technologies, skilled labour, and the iterative nature of the design process can all contribute to higher expenses compared to traditional manufacturing methods.
- Manufacturing Tolerances: Discrepancies between digital models and physical products can occur due to manufacturing tolerances and material behaviours. Ensuring that the final product matches the design specifications requires tight quality control and may involve additional steps such as post-processing or adjustments during fabrication.
- Scalability: While parametric design excels in creating customized solutions, scaling up production for larger projects or mass production can be challenging. The unique nature of each design iteration may not lend itself well to efficient mass production, where standardization and repeatability are key.

The novel research approach based on no-mould processes aims to lower costs and increase the sustainability of free-form panel production by reducing assembly waste [31]. Two technologies that are entering the market with good potential are the "cement textile composite" (CTC) [32] and "adaptive moulding" [33].

The two technologies are based on different principles. The "adaptive moulding" system integrates into an existing supply chain of thermoformable materials, utilizing a deformable membrane manipulated by hydraulic jacks. This system avoids the creation of single-use moulds, thereby speeding up the process; however, it requires significant initial investments for the development of the machinery. On the other hand, the CTC solution uses a stretchable fabric and a frame to create rigid surfaces with varying degrees of customization, stiffness, and transparency, without the need for moulds. This makes it applicable with modest investments and accessible to multiple manufacturers and craftsmen making it more applicable at different industrial scales.

5.4.1 Unlocking Innovation: The Parametric Approach in Cement-Textile Composite (CTC) Technology

CTC is a new composite material that combines a synthetic reinforcing textile with a customized concrete matrix. It has been developed to generate rigid elements capable of preserving complex spatial arrangements, particularly double-curvature surfaces. This innovative material is produced using a 3D-warp knitted polyester fabric combined with a tailor-made cement mix. The technology is based on a cementitious matrix (cement, water, and polymer fluid) coupled with a deformable three-dimensional fabric. The realization of these surfaces does not require formworks, moulds, or special casts, allowing for a quick and cost-effective forming process. The selection of a 3D fabric is crucial in achieving the target shape; inadequate elasticity may require implementing joining lines, determined through an optimization process.

This technology preserves the finishes of the fabrics used, allowing for different types of cement (white, traditional, coloured, high-strength, or low environmental impact) to be employed. The treatments and production processes preserve the geometric texture, colour, or pattern of the fabric and its tactile sensation without additional processing after installation. This developed technology also offers the possibility to generate cladding surfaces with complex continuous/discontinuous or semipermeable geometries (e.g., for light or air).

The shape can be achieved by stretching the fabric over a frame (made using traditional processes in wood, aluminium, etc.) on which it is then fixed. Subsequently, the fabric is impregnated with a cement mixture which, upon hardening, will make the surface rigid as presented in Fig. 5.1. Alternatively, the production process may involve first impregnating the fabrics and then forming them through deformation by localized and/or distributed compression and tension actions. The mixture allows for

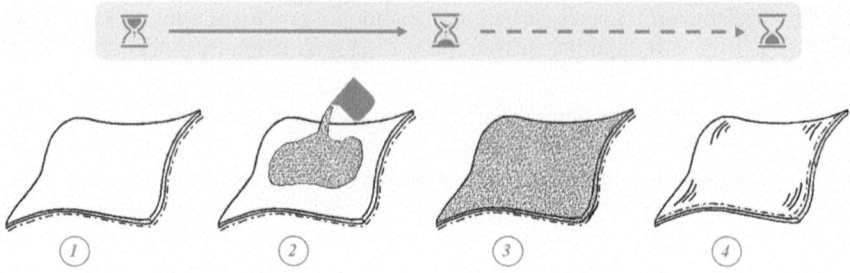

Fig. 5.1 CTC realization process

about an hour of workability with a setting time of 2–6 h, enhancing the efficiency of the production process. Various parameters can be controlled, such as fabric thickness (between 3–20 mm), texture, and colouration (white is standard, but any colour can be chosen, and images or designs can also be printed). The dimensions, mechanical properties, and many other characteristics depend on the fabric manufacturer. The cement matrix's rheological behaviour has been customized for specific fluidity and homogeneity according to fabric characteristics, such as the openness factor of the front surface and the space between interlayer yarns. Fluidity and filler dimensions can be easily adjusted for different configurations while ensuring sufficient workability, stiffness, and rigidity.

The technology developed allows the preservation of the fabric finishes employed, utilizing different types of cement (white, traditional, coloured, high-strength, or low environmental impact). The production processes ensure that the geometric texture, colour, or pattern of the fabric, as well as its tactile sensation, are preserved without additional processing after installation.

This material, therefore, allows for the creation of complex surfaces through simple processes, delegating the management of complex variables to the initial engineering phase (Fig. 5.2). As a result, the production of artefacts based on this technology can be achieved with varying levels of production machinery development, ranging from a basic 3-axis work centre for the frame and a laser cutter for the fabric to 3D printers for aluminium supported by robotic arms. Another significant advantage is the ability to use multiple materials for the frame, the tensioned membrane (various types of fabric), and the stiffening matrix. This flexibility enables production in different parts of the world without the necessity for specific machinery that may be more common in certain areas or materials that are only available in specific regions.

Another advantage is the high level of customization. Both the colouration and transparency of the fabric can be managed. This allows for the colouring processes to be applied directly to the fabric (which will be protected), eliminating the need for post-forming colouring processes that would require more effort and cost, especially for a complex surface. Furthermore, the possibility of creating translucent components by adjusting the cement mix or removing it locally, makes possible the control of the translucency of the element, creating specific light patterns. The hybrid nature

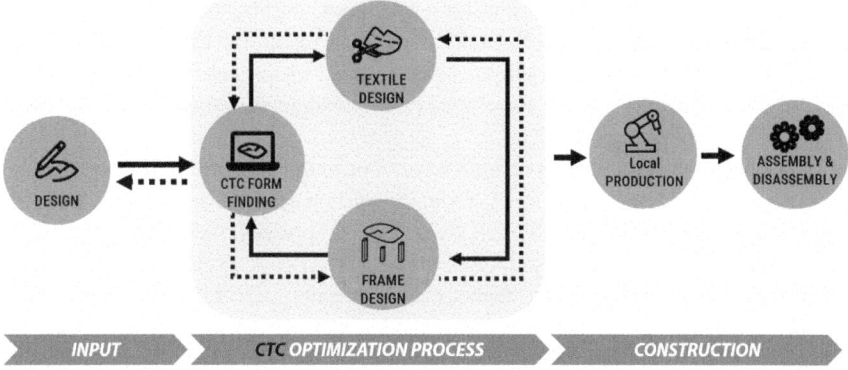

Fig. 5.2 CTC design phases

of CTC, blending fabric characteristics like texture, translucency, and ductility with thin-shell tectonics, represents cutting-edge technology for developing free-form building envelope components, such as high-performance solar shading systems.

Based on the previous description, it is possible to say that CTC is a production technology that leverages morphing membrane principles, signifying a surface-based modelling approach, in contrast to the volume-based methodologies employed, for example, by CNC and FDM 3D-printing technologies. For this reason, it is possible to infer that working with membranes is highly advantageous in terms of material usage, especially when the form has a three-dimensional development. Comparing different production technologies for the same shape results in a significant carbon saving—up to 90% of the carbon footprint—when CTC technology is adopted [34].

5.4.2 Parametric Performance in Building Envelope: Form Energy to Well-Being-Driven Approach

The significance of studying parametric building envelopes lies in their potential to enhance energy efficiency, optimize thermal performance, and support sustainable building practices. Parametric studies allow for the examination and optimization of various design configurations, which can lead to improved energy performance and reduced operational costs. For example, the research by Kalús et al. [35] investigates the energy potential of building envelopes with integrated energy-active elements, showing that specific configurations can minimize heat loss and improve energy efficiency. Zhang et al. [36] propose a parametric model for evaluating the thermal performance of building envelopes containing phase change materials, facilitating quick and cost-effective assessments. Han's [7] study on high-rise residential build-ings in India demonstrates how parametric optimization can significantly reduce cooling demand and operational energy use while Mučková et al. [37] explore the dynamic thermal barriers in building envelopes, highlighting the potential for substantial improvements in thermal resistance and insulation efficiency. The literature review by Pour et al. [38] emphasizes the need for probabilistic hygrothermal assessments to ensure building envelopes can withstand varied climatic conditions while maintaining performance.

The primary objectives of an energy-driven approach in building envelope design include optimizing energy efficiency, reducing carbon emissions, and enhancing the overall sustainability of the built environment. This approach focuses on integrating energy considerations into every stage of the design process, from conceptual design to detailed architectural planning. Vafaee et al. [39] emphasize the importance of integrating energy considerations during the early design stages to optimize building form and reduce energy consumption through parametric design tools and environmental simulations. Huang et al. [40] discuss using deep reinforcement learning to automate urban design, aiming to minimize energy use and maximize renewable energy potential. This energy-driven approach often employs advanced computational methods and data-driven techniques to achieve precise and effective design solutions. Wang et al. [41] propose an optimization framework that incorporates urban morphology to maximize solar energy utilization and minimize building energy demand, demonstrating how energy-driven design can significantly impact urban sustainability. The goal is to create buildings and urban environments that are not only energy-efficient but also contribute to broader environmental objectives such as carbon neutrality and resilience to climate change.

The key performance indicators (KPIs) for energy efficiency in parametric building envelopes focus on metrics that help quantify and optimize energy use, thermal performance, and environmental impact. These KPIs are essential for assessing the effectiveness of building envelope designs in reducing energy consumption and enhancing sustainability. Here are some primary KPIs:

- Transmission Losses measure the heat transfer through the building envelope, including walls, roofs, windows, and doors. Lower transmission losses indicate better insulation and energy efficiency.
- Heating and Cooling Energy Consumption tracks the energy required for maintaining indoor thermal comfort. Effective building envelopes minimize heating and cooling demands by providing good insulation and preventing unwanted heat gains or losses.
- Greenhouse Gas Emissions (GHGs) are a critical KPI, as they reflect the building's environmental impact. Energy-efficient building envelopes contribute to lower emissions by reducing energy consumption from non-renewable sources.
- Thermal Comfort assesses the indoor environment's ability to maintain comfortable temperature and humidity levels without excessive reliance on mechanical systems.
- Financial Costs for Building Maintenance as reducing the operational costs associated with heating, cooling, and maintenance, contributes to long-term financial savings.

- Passive Cooling Techniques are strategies such as natural ventilation, shading devices, and reflective materials to reduce cooling loads and improve energy efficiency.

In parallel with the approach applied in the context of parametric optimization of energy performance, the concept of occupant well-being has also gained prominence in recent years. This concept, considering multiple parameters, is beginning to be regarded as a key point for parametric modelling. It is further supported by the incorporation of these concepts into evaluation protocols such as LEED [42] and WELL [43].

User well-being significantly influences the design and performance of building envelopes by dictating the requirements for comfort, health, and satisfaction within indoor environments. Various studies illustrate how user well-being shapes building envelope design. Vladoiu et al. [44] highlight that ensuring indoor comfort is paramount in the design of energy-efficient buildings. Factors such as hygrothermal, visual, and acoustic comfort directly impact the quality of the indoor environment and the well-being of occupants. Therefore, building envelopes must be designed to control these parameters effectively, meeting specific design norms and standards to enhance user well-being.

Furthermore, de Paiva [45] emphasizes the role of neuroscience in architecture, noting that understanding how space affects brain functions can help architects design buildings that improve user behaviour, performance, and overall well-being. Building envelopes that consider the psychological impact of spaces can foster social relations, focus, cognition, creativity, and mental health, contributing to a more holistic approach to design.

Luna-Navarro and Overend [46] propose that smart building envelopes, which can adapt to changing outdoor conditions and indoor requirements, offer a low-carbon means of achieving occupant satisfaction and well-being. These adaptive envelopes enhance comfort by regulating energy transfer and maintaining optimal environmental conditions, thereby supporting user health and productivity.

Watson [47] discusses the need for evidence-based design that delivers well-being outcomes. Establishing metrics for psychological well-being in the built environment can help designers create user-centred spaces that enhance mental health and overall satisfaction. This includes considering factors such as natural light, air quality, and biophilic design elements.

User well-being in building design includes several key aspects that ensure the physical, psychological, and emotional health of occupants. These aspects are:

- *Thermal Comfort*: This refers to the ability of the indoor environment to maintain temperatures within a range that is comfortable for occupants. Effective thermal comfort minimizes the need for artificial heating and cooling, thereby promoting energy efficiency and occupant satisfaction.
- *Visual Comfort*: Visual comfort involves providing adequate natural and artificial lighting that reduces glare and ensures sufficient illumination for various activities.
- *Air Quality*: High indoor air quality is crucial for health and well-being, involving proper ventilation, control of pollutants, and maintaining appropriate humidity

levels. Good air quality helps prevent respiratory issues and contributes to overall comfort.

- *Acoustic Comfort*: This involves managing noise levels within buildings to ensure a quiet and peaceful environment. Effective acoustic design includes sound insulation and absorption techniques to reduce noise pollution from both internal and external sources.
- *Safety and Accessibility*: Ensuring that buildings are safe and easily accessible for all users is fundamental to well-being. This includes designing for physical safety, providing clear egress routes, and incorporating universal design principles for accessibility.
- *Psychological and Emotional well-being*: Architectural design can significantly impact mental health by creating environments that promote relaxation, reduce stress, and enhance mood. Elements such as natural views, biophilic design, and aesthetically pleasing spaces contribute to psychological well-being.
- *Ergonomics and Functionality*: Designing spaces that are functional and ergonomically sound supports the physical health of occupants. This includes appropriate furniture design, spatial layout, and the overall usability of the space.
- *Social Interaction and Privacy*: Balancing spaces for social interaction with areas for privacy helps meet the social and emotional needs of occupants. This includes designing communal areas as well as private spaces where individuals can retreat and relax.

As described energy-driven and user well-being approaches in the context of parametric building envelopes serve distinct yet complementary roles in enhancing building performance. Energy-driven approaches focus primarily on optimizing the energy efficiency of building envelopes to reduce operational energy consumption and improve overall sustainability. On the other hand, user well-being approaches prioritize the comfort and health of building occupants, ensuring that indoor environments maintain optimal conditions for human well-being. Both approaches intersect in their shared goal of creating high-performance buildings. Therefore, a holistic approach that combines energy efficiency with user well-being can lead to more sustainable and liveable buildings. Within the concept of well-being, the need to enhance and maximize the view out is gaining increasing importance, especially following the experience of the lockdown during the COVID-19 period.

5.4.3 View Out as a New Comfort Parameter for Occupant Well-Being

Visual perceptions acquired from a certain environment can significantly influence the sense of well-being experienced by the occupants of that space. This seemingly simple correlation, which many of us may have often considered throughout our lives, is actually a complex topic with vast implications and a wide range of interconnected issues.

In recent years, research on this topic has seen considerable growth. As Rodríguez Iturriaga commented [48], the massive social experiment of the COVID-19 lockdowns created an atmosphere of global reflection on contemporary urban landscapes and, more broadly, on how building users perceive, interpret, and value their surrounding environment. Discussions about the future of architectural design and urban planning quickly focused on what we might refer to as the "trending topics" of that time: "virtual space" and "outdoor space" [49].

Following those events, the topic of "outdoor space" and the ability to perceive it were explosively rediscovered as key factors in the well-being of habitable spaces. As already mentioned, such issues were part of previous discussions. Note, for example, declared in different circumstances how most of the buildings we live in have become so efficient at protecting occupants from the elements that they also eliminate some key requirements for our well-being: nature and change [52]. Isolation and "poor" visual conditions may decrease work performance [53] or increase the stress levels of the occupants. However, it is not easy to solve these complexities because of the high difficulty of framing the issue itself, alongside the challenge of defining the boundaries for investigating a certain space, and finally the metrics for evaluating possible solutions. Setting the boundaries for evaluating the visual conditions of a living environment requires defining the value of perceptual inputs across space. Setting, for example, a standard home as a basic case study, Cullen [54] argues that even such a basic spatial construct can be considered a micro-landscape of architectural scale that filters and introduces fragments of the larger landscape outside. As a result, the habitable space of a home is not severed from "outside space" at the boundaries of the building itself but, with varying degrees of effectiveness, virtually extends to the sky, the horizon, and the surrounding areas, up to the limits of multisensory perception.

Therefore, dealing with the visual aspects of a built environment requires the application of a multi-scale approach that enables us to grasp the inner complexity of these phenomena.

For this reason, it can be said that research on outdoor views aims to improve the general well-being derived from using a given built environment by optimizing the design of visual interactions with the outside space. To assist in achieving this goal, specific evaluation targets, data acquisition procedures, and evaluation metrics are developed accordingly [55, 56]. Some examples of such aids can be found in the numerous building certifications and guidelines that include specific evaluation systems to promote the presence of "quality views" within buildings as presented in Table 5.3.

An analysis conducted within an ongoing research study spearheaded by SEEDLab @DABC allowed for the testing of the workflow for visual coverage calculation within a multi-criteria analysis framework. This resulted in combined analyses of view and light quality [57]. The goal of the study was to evaluate the impacts of residential parametric energy retrofits, focused on perimeter walls, on the ability of the windows within them to transmit both light and views (Fig. 5.3).

Different test environments were carefully constructed to generate generalized scenarios regarding Italian building typologies and dwelling settings. Finally,

Table 5.3 Comparison between the most established evaluation guidelines concerning the outdoor view within buildings

Certification/guidelines	Parameter/credit	Year	Scale	References
LEED v4.1	Quality view(s)	2020	Room, building	[42]
WELL v2-pilot	View factor	2020	Room, building	[43]
European standard EN 17,037 on daylighting	Level of view quality	2018	Room, building	[50]
Society of Light and Lighting, Lighting Guide (SLL-LG)	Level of view quality	2014	Room, building	[51]

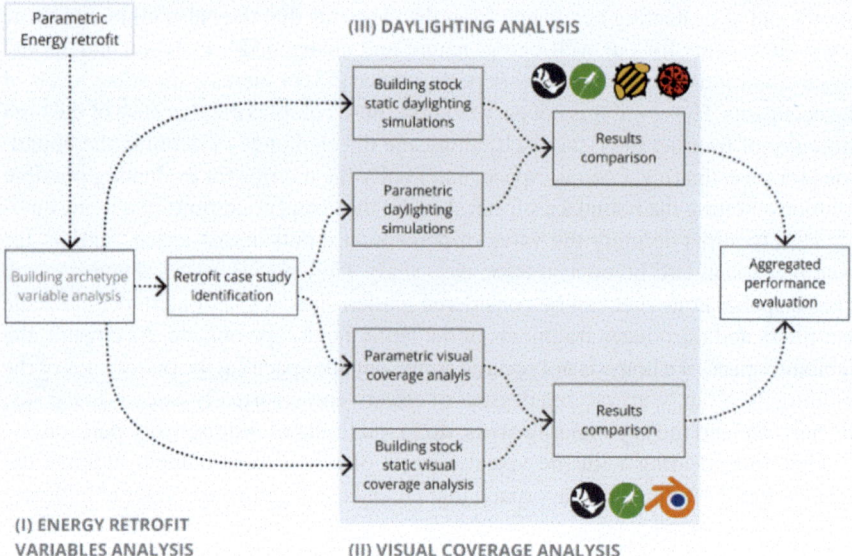

Fig. 5.3 The workflow defined within the multi-criteria analyses on light and view was carried out as part of the SEEDLab @DABC experimentation framework

different retrofit scenarios were simulated by defining three new sets of test environments, each built upon the base group through the application of a specific retrofit strategy.

The conditions regarding daylight in any of the resulting environments were evaluated through analyses of the Daylight Factor (DF), the Daylight Autonomy (DA), and the Useful Daylight Illuminance (UDI). Meanwhile, the visual analyses based on Visual Coverage (VC) were calculated concerning the generic "external view," without specifying its actual contents or characteristics. In this way, the VC is used to generically extract basic information about the "amount" of outdoor view in the environment.

A comparison between the DF and the VC highlights similar variations between the two readings regarding the application of parametric retrofit strategies focused

on the walls. While this is due to a common dependency between the two calculation methodologies on the spatial property of "obstruction," it is important to note
that while the DF is associated with multiple evaluation guidelines, as previously
mentioned, the same is not true for the outdoor view evaluation as a whole. Therefore,
similar experiments, which essentially link a well-studied phenomenon with a less-
known one, may ultimately allow further development of outdoor view evaluation
through knowledge transfer from related fields.

5.5   Unlocking the Parametric Building Envelope

The increasing level of complexity and the different needs arising from multiple
crucial aspects are essential for the development of building envelopes. In this context,
a paradigm shift is underway to bridge detailed, human-centric building and envelope
design with their operation (Fig. 5.4).

This chapter explores how the concept of parametric design is currently applied to
the building envelope sector. There are various perspectives on parametric modelling

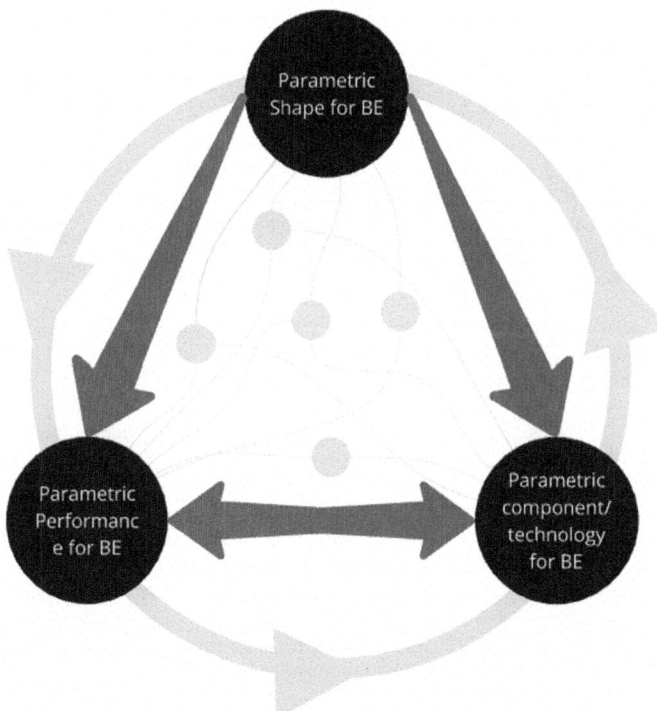

Fig. 5.4 Circular multi-parameter optimization approach schematization

of the envelope, each of which can have varying importance depending on the application context. Three perspectives that have been presented are: parametric shape, supported by parametric components/technologies, and parametric performance.

When applied to research and development projects within the building envelope domain, these three aspects often exhibit strong integration rather than distinct separation, as they are closely interconnected realities. Currently, the most widely applied approach is a linear and one-directional method, where starting from the most prevalent aspect, the secondary aspect is defined retrospectively.

To enable a genuine unlocking of the potential of parametric analyses applied to the building envelope, it is important to transition from the current approach, which is based on the optimization of a single parameter, to a circular, multi-parameter optimization approach that equally weighs different drivers. This circular approach allows for the enhancement of multiple aspects concurrently. Moreover, this optimization should not only be multi-parameter (considering the individual variables within a domain) but also multi-domain, thus enhancing the overall parametric approach comprehensively.

References

1. Khamis AA, Ibrahim SA, Khateb MA et al (2022) Introducing the architecture parametric design procedure: from concept to execution. IOP Conf Ser Earth Environ Sci 1056:012004. https://doi.org/10.1088/1755-1315/1056/1/012004
2. Shi B, Hao K (2022) Research on the application of the parametric design method in special-shaped buildings. Advances in urban engineering and management science, vol 1. CRC Press, London, pp 363–368
3. Moreira F, Nogueira A, Santos H et al (2023) Eco parametric architecture: circular design and digital fabrication
4. Niphadkar PA, Niphadkar AP (2022) Strategies for parametric design in architecture. Int J Adv Res Sci Commun Technol 115–119. https://doi.org/10.48175/IJARSCT-3774
5. Abdollahi Rizi R, Sangin H, Haghighatnejad Chobari K et al (2023) Optimising daylight and ventilation performance: a building envelope design methodology. Buildings 13:2840. https://doi.org/10.3390/buildings13112840
6. Manavis A, Firtikiadis L, Spahiu T, et al (2022) Parametric architectural design using shapes and structures. J Graph Eng Des 13:13–20. https://doi.org/10.24867/JGED-2022-4-013
7. Chaturvedi PK, Kumar N, Lamba R, Nirwal V (2023) A parametric optimization for decision making of building envelope design: a case study of high-rise residential building in Jaipur (India), pp 453–465
8. Mendilcioğlu RF, Yılmaz M (2021) Parametric design as a creation tool for the memory space. Iconarp Int J Archit Plan 9:429–454. https://doi.org/10.15320/ICONARP.2021.166
9. Brakke AP, Velasco R (2013) Visualizing the expressive use of technology in the design of parametrically generated eco-envelopes
10. Hou D, Liu G, Zhang Q et al (2017) Integrated building envelope design process combining parametric modelling and multi-objective optimization. Trans Tianjin Univ 23:138–146. https://doi.org/10.1007/s12209-016-0022-1
11. Lin Y, Shen S (2012) Designing a performance-oriented house envelope based on a parametric approach: an integrated method, pp 507–516
12. Steinø N (2019) Parametric urban design from concept to practice. Blucher Des Proc
13. Union E New European Bauhaus. https://new-european-bauhaus.europa.eu/index_en

14. Piroozfar P, Piller F (2013) Mass customisation and personalisation in architecture and construction. Routledge
15. Gawell E (2014) Non-euclidean geometry in the modeling of contemporary architectural forms
16. Álvarez Elipe MD, Anaya Díaz J (2018) Review of contemporary architecture projects based on nature geometries. Rev la construcción 215–221. https://doi.org/10.7764/RDLC.17.2.215
17. Ostwald M (2012) Systems and enablers: modeling the impact of contemporary computational methods and technologies on the design process 1–17. https://doi.org/10.4018/978-1-61350-180-1.ch001
18. Toromanoff A (2021) Curved: bending architecture. Lannoo
19. Kim K, Son K, Kim E-D, Kim S (2015) Current trends and future directions of free-form building technology. Archit Sci Rev 58:230–243. https://doi.org/10.1080/00038628.2014.927751
20. Brookes AJ, Brookes MM (2008) Cladding of buildings, 4th Edn
21. Souza E. (2022) What materials can be used for façade cladding? https://www.archdaily.com/979639/what-materials-can-be-used-for-facade-cladding?ad_source=search&ad_medium=search_result_articles. Accessed 29 Jun 2023
22. Marius M, Anastasiadis A, Kampouris A (2014) Are free form architecture ecological buildings. J Environ Prot Ecol
23. Kavuma A, Ock J, Jang H (2019) Factors influencing time and cost overruns on freeform construction projects. KSCE J Civ Eng 23:1442–1450. https://doi.org/10.1007/s12205-019-0447-x
24. Hambleton D, Howes C, Hendricks J et al (2009) Study of Panelization techniques to inform freeform architecture
25. Ock J-H (2021) Testing as-built quality of free-form panels: lessons learned from a case study and mock-up panel tests. Appl Sci 11:1439. https://doi.org/10.3390/app11041439
26. Moskaleva A, Safonov A, Hernández-Montes E (2021) Fiber-reinforced polymers in freeform structures: a review. Buildings 11:481. https://doi.org/10.3390/buildings11100481
27. Hom Ock J (2018) Assessing the suitability of the cold bending method in fabricating free-form Façade Panels. Civ Eng Res J 5. https://doi.org/10.19080/CERJ.2018.05.555653
28. Alonso-Pastor L, Lauret-Aguirregabiria B, Castañeda-Vergara E et al (2014) Free-form architectural façade panels: an overview of available mass-production methods for free-form external envelopes. In: Llinares-Millán C, Fernández-Plazaola I, Hidalgo-Delgado F et al (eds) Construction and Building Research. Springer, Netherlands, Dordrecht, pp 149–156
29. Castañeda E, Lauret B, Lirola JM, Ovando G (2015) Free-form architectural envelopes: digital processes opportunities of industrial production at a reasonable price. J Facade Des Eng 3:1–13. https://doi.org/10.3233/FDE-150031
30. Han D, Yin H, Qu M et al (2020) Technical analysis and comparison of formwork-making methods for customized prefabricated buildings: 3D printing and conventional methods. J Archit Eng 26. https://doi.org/10.1061/(ASCE)AE.1943-5568.0000397
31. Jipa A, Bernhard M, Dillenburger B et al (2017) Skelethon formwork 3D printed plastic formwork for load-bearing concrete structures. In: Blucher design proceedings. Editora Blucher, São Paulo, pp 345–352
32. Poli T, Zani A, Speroni A, Mainini AG (2019) Cladding element for use in construction and method for manufacturing the same
33. Adapa adaptive Moulds. https://adapamoulds.com/. Accessed 26 Jul 2024
34. Speroni A, Cavaglià M, Mainini AG et al (2023) Parametric assessment to evaluate and compare the carbon footprint of diverse manufacturing processes for building complex surfaces. Buildings 13. https://doi.org/10.3390/buildings13122989
35. Kalús D, Koudelková D, Mučková V et al (2022) Parametric study of the energy potential of a building's envelope with integrated energy-active elements. Acta Polytech 62:595–606. https://doi.org/10.14311/AP.2022.62.0595
36. Zhang Y, Jiang W, Song J et al (2023) A parametric model on thermal evaluation of building envelopes containing phase change material. Appl Energy 331:120471. https://doi.org/10.1016/j.apenergy.2022.120471

37. Mučková V, Kalús D, Koudelková D et al (2023) Analysis of the dynamic thermal barrier in building envelopes. Coatings 13:648. https://doi.org/10.3390/coatings13030648
38. Bayat Pour M, Niklewski J, Naghibi A, Frühwald Hansson E (2024) A literature review of probabilistic hygrothermal assessment for building envelopes. Build Environ 261:111756. https://doi.org/10.1016/j.buildenv.2024.111756
39. Vafaee NZ, Sandani M, Khameneh TA et al (2022) Integrating energy in the conceptual design stage to optimize building form, pp 318–324
40. Huang C, Zhang G, Yin M, Yao J (2022) Energy-driven intelligent generative urban design, based on deep reinforcement learning method with a nested deep Q-R network, pp 233–242
41. Wang W, Liu K, Zhang M et al (2021) From simulation to data-driven approach: a framework of integrating urban morphology to low-energy urban design. Renew Energy 179:2016–2035. https://doi.org/10.1016/j.renene.2021.08.024
42. USGBC (2019) LEED v4 CREDITS for building design and construction. LEED Publ 147
43. International WELL Building Institute (2020) WELL building standard v2
44. Vladoiu CL, Isopescu DN, Maxineasa SG (2021) Indoor environment from wellbeing perspectives, pp 67–88
45. de Paiva A (2018) Neuroscience for architecture: how building design can influence behaviors and performance. J Civ Eng Archit 12. https://doi.org/10.17265/1934-7359/2018.02.007
46. Luna-Navarro A, Overend M (2018) Towards human-centred intelligent envelopes: A framework for capturing the holistic effect of smart façades on occupant comfort and satisfaction. Healthy, intelligent and resilient buildings and urban environments. International Association of Building Physics (IABP), Syracuse, New York, pp 661–666
47. Watson KJ (2018) Establishing psychological wellbeing metrics for the built environment. Build Serv Eng Res Technol 39:232–243. https://doi.org/10.1177/0143624418754497
48. Rodríguez Iturriaga M (2021) Learning from COVID-19: the role of architecture in the experience of urban landscapes. Ri-Vista Res Landsc Archit 19:122–137. https://doi.org/10.36253/rv-10182
49. Nasrollahzadeh Mehrabadi E, Pilehchi Ha P, Mahdavinejad M (2021) Horsefly: a simulation tool to evaluate view to outdoor
50. BSI Standards Publication (2018) BS EN 17037: daylight in buildings
51. CIBSE (2014) LG10 daylighting—a guide for designers 92
52. Nute K, Weiss A (2016) Outside in: using the animation of the weather to improve building occupants' well-being and raise awareness of passive energy and rainwater saving. Int J Architecton Spat Environ Des 10:41–56. https://doi.org/10.18848/2325-1662/CGP/v10i04/41-56
53. Ko WH, Schiavon S, Zhang H et al (2020) The impact of a view from a window on thermal comfort, emotion, and cognitive performance. Build Environ 175:106779. https://doi.org/10.1016/j.buildenv.2020.106779
54. Cullen G El paisaje urbano: tratado de estética urbanistica
55. Li W, Samuelson H (2020) A new method for visualizing and evaluating views in architectural design. Dev Built Environ 1:100005. https://doi.org/10.1016/j.dibe.2020.100005
56. Matusiak BS, Klöckner CA (2016) How we evaluate the view out through the window. Archit Sci Rev 59:203–211. https://doi.org/10.1080/00038628.2015.1032879
57. Mainini AG, Cavaglià M, Blanco Cadena JD, Speroni A, Poli TM (2023) Residential energy retrofits: balancing daylight performance and outdoor view

Chapter 6
Investigating Decision-Making Frameworks for Early-Stage Performance-Based Building Envelope Design

Abstract This chapter explores the critical role of early-stage design in the architectural process, emphasizing the importance of managing the building envelope to enhance overall building value and occupant well-being. It introduces a generalized graph database designed to assist in decision-making by mapping and navigating the complex relationships between various design factors and performance domains. The chapter demonstrates how a graph database can serve as a valuable tool for structuring search queries, ranking design variables, and providing a comprehensive overview of interdependencies. Future research directions include integrating this data with parametric software systems and exploring web-based solutions for enhanced accessibility. Ultimately, this approach aims to support the development of effective design strategies throughout the design process, enabling the extraction of generalized outlooks of complex design relationships and fostering a holistic view of building design.

6.1 Introduction

The previous chapters emphasized, in various instances, the deep complexity determined by the interconnected nature of building components. Historically, schemas and infographics have been crucial in describing, and visually mapping, this intricate network of relationships within the multi-domain description of the building design elements. An example of this can be seen in Fig. 6.1, where a diagram expanded from Fitch [1], is here used to easily visually comprehend the many connections intermediated between the internal and external environments by the building envelope.

These kinds of outputs can be regarded as the endpoint visualization of previously collected data, meaning, schema and infographics can be viewed as the final output resulting from the interaction with a specific database [2]. For a long time, pictures and databases have rarely been dynamically interconnected, making it challenging to effectively pair them as tools in the decision-making process during the design stage. However, due to the clear advantages of such dynamic interactions, automation of

© The Author(s), under exclusive license to Springer Nature Switzerland AG 2024 103
A. G. Mainini et al., *Unlocking the Potential of Building Envelopes*,
PoliMI SpringerBriefs, https://doi.org/10.1007/978-3-031-75298-8_6

Fig. 6.1 Adapted and expanded from Fitch [1]. While the original image from Fitch generally described the role of "walls" in building design (effectively implicitly considering windows as part of the walls regarding the schema compilation), in this context the data has been reframed and expanded to the concept of building envelope design

data extraction procedures has become widely implemented to visually structure the complexity of potential results. This concept is widely applied in specific segments of building design, where "product configurators" visually display the output of a parametric process able to sort a database to provide useful information for integrating specific systems or components into the design. Similarly, managing multi-domain dependencies at the building level can unlock a more advanced understanding of the impact and reverberation of design strategies, particularly during early-stage design, as discussed in Chap. 2.

6.2   Harnessing Graph Database Technology for Improved Data-Driven Building Envelope Design

Effectively managing the dependency network among the relevant factors in a design essentially involves the definition of a workflow suitable for handling information which, in turn, necessitates the creation of a database. Storing the relevant information within a proper database also means to enable the integration of the stored data within multiple external environments, while also leveraging interactive and visualization applications to seamlessly link the implementation of the information into the decision-making processes.

In this regard, a graph database can be seen as an adequate fit for the required task. A graph database is a type of database which stores data as a network graph [3] and it is fundamentally composed by two elements: nodes and edges. Nodes serve as the primary entities in the graph. These can represent any item that the database needs to track. Furthermore, nodes can have a custom number of attributes which may represent additional information about each specific node [4].

On the other hand, edges, also commonly referred to as relationships or links, define the connections between the nodes. Each edge connects two nodes, creating a link that can be elaborated. At its core, an edge signifies a relationship between two nodes (e.g., items in a database). Additional information can be added using "directed edges", which prioritize an "ordered" relationship, establishing a basic binary hierarchy with one element preceding the other. Edges might also include attributes to provide additional details about the nature of the relationship between the nodes, just as the nodes themselves [4]. Figure 6.2 presents a summary of the presented concepts.

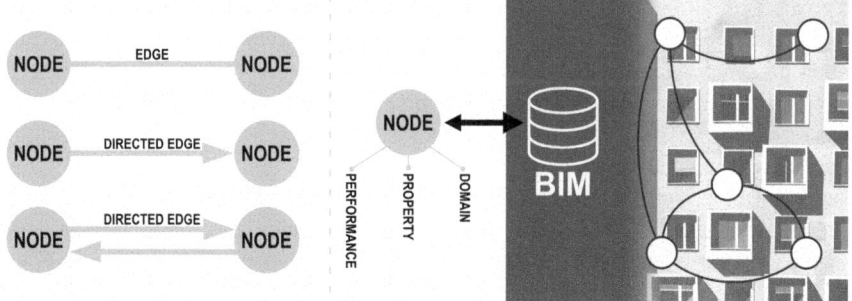

Fig. 6.2 The picture displays the basic structure of a graph database, namely the interconnection of two nodes via an edge. The picture further details the possible state for an edge relationship: undirected (above) and directed (below). A graph database can be used to store multiple dataset and may be also implemented in the BIM environment

The structure of a graph database is particularly well-suited to clearly and visually represent the connections between multiple entities with intricated relationships, unlike other database models that primarily manage data via tables (e.g. relational databases) [5]. Every aspect of the built environment can inherently be deconstructed into a series of relationships among various elements of the design, whether they are material or immaterial. These relationships can encompass spatial, functional, temporal, socio-economic, environmental, and architectural aspects. Therefore, a graph database can effectively map relevant data to assist in decision-making processes, providing a clear summary of complex issues through customizable and interactive visualization tools.

Numerous platforms exist for managing graph databases, varying in their level of external accessibility. Some platforms operate in a closed environment, focusing mainly on internal graph creation, management, and graphical exports. Alternatively, other platforms offer real-time data access to different software environments, allowing specific data retrieval for external implementation [6]. In any case, although the choice of the software environment to implement may hugely affect the overall capability of extensively implementing the resulting graph database, it is also important to note that appropriate file formats allow for efficient import/export operations, allowing the complete migration of a given database towards different environment if needed. Among the list, it is possible to cite as an example the formats, graphML (.graphml), GEXF (.gexf), and GraphJSON (.json) [7]. In this context, the open-source software Gephi [8] has been selected as the environment to produce a dependency network in the framework of building envelope optimization and design.

This database is envisioned to function as a tool to aid decision-making strategies, while additionally serving as a straightforward visualization tool for examining the potential ramifications of specific design instances. In particular, a comprehensive list of design variables coupled with data about their underlying connection is regarded to be a potential answer to the limit of parametric design optimization emphasised in Chap. 2, Sect. 2.2. The advantages of such a database are entirely defined by the extent of its application, but in this context, it is regarded that the primary aim is to offer a detailed, visual, and interactive map of design feature interdependencies. This objective, in and of itself, could serve as a valuable platform for formulating decision-making strategies.

Within the context of building envelope design, the database follows a primary hierarchical structure defined from a main level divided into three fundamental macro-domains. These domains serve as roots for features related to the technological components, environmental factors, and people-related aspects. A technological component refers to any element or system that leverages technology to enhance the design, construction, or functionality of a building [9], the following database has primarily addressed the technological component of the transparent envelope, as a possible core unit in building envelope design, and only for this node, a further root path is detailed to highlight the possibility of expanding the database scope beyond the proposed targets.

Moving on, the environment is here broadly considered as the spatial area related to the project design, this obviously addresses the building site and living space of said project, but it could also extend to its perceptive surroundings. As Fitch noted, architecture is entirely enveloped by its external surroundings. As such, it's impossible to truly grasp, observe, or experience it without acknowledging its multi-dimensional entirety [1]. A concept that Iturriaga summed up in the suggestion that the actual "lived space" can "*virtually extend to the sky, the horizon, and the surrounding areas, up to the limits of multisensory perception* [10]".

The last macro-area of analysis is here labelled as "people", and it is aimed to store and address any design feature connected to the various aspects that the interaction between people and the built environment might take on. This may account for both the people's needs that must be addressed within the project, and the various instances of behaviours that people may act as inhabitants in the living space.

In Fig. 6.3 it is possible to observe the further explosion of the main level of the database, into the subsequent sublevels. More specifically, at the first sublevel, a second hierarchy of categories is utilized to further organize the subsequent entities within the database. Lastly, a third sublevel is used to sort any kind of KPI related to the second sublevel nodes.

The selection process which sorted and finally composed each list is the result of multiple cross-examination based on assessments about the design practice and reviews of correlated topics.

To recap the foremost characteristics of the resulting database. The technological component group, which in this context is focused on the transparent envelope, contains various characteristics that can be used to define its design. Similarly, the environment group stores different characteristics which may be used to define an environment. In this regard, the most direct sublevel describes the features which may be used in detailing the layout structure of the space. Also, in addition to the layout features, two other groups have been opted to define other subgroups on the same level: the design domain and the performance domain. Both of these groups address distinct aspects the design which overall may be implemented to evaluate and guide the design of the built environment. Finally, the people node extends into two sublevels, labelled as: "needs" and "behaviour". As the names suggest, each sublevel deals respectively with the specific needs of the users, and the behaviours may enact in the space. However, to better represent different nuances of said features, the elements picked to populate these groups were additionally categorized in even more detail. People's behaviours, for example, are distributed into two categories: instantaneous behaviours and persistent behaviours. Instantaneous behaviours encompass actions that can be executed at high frequency and with little preliminary planning operations, such as opening a window. On the other hand, persistent behaviours address actions that have low frequency and require more complex planning, such as changing the layout of the pieces of furniture. Figure 6.4 displays the endpoint of this process.

At the stage displayed in Fig. 6.4, the database now contains all the most relevant design factors which may play an important role when dealing with the design of the transparent envelope. While not immediately visible, the third sublevel defined to store performance indicators linked to each performance domain recorded is also accounted for in the form of a list object. Figure 6.5 shows a linked table, which at the present state acts as a simple referenced object to store a related KPI-list.

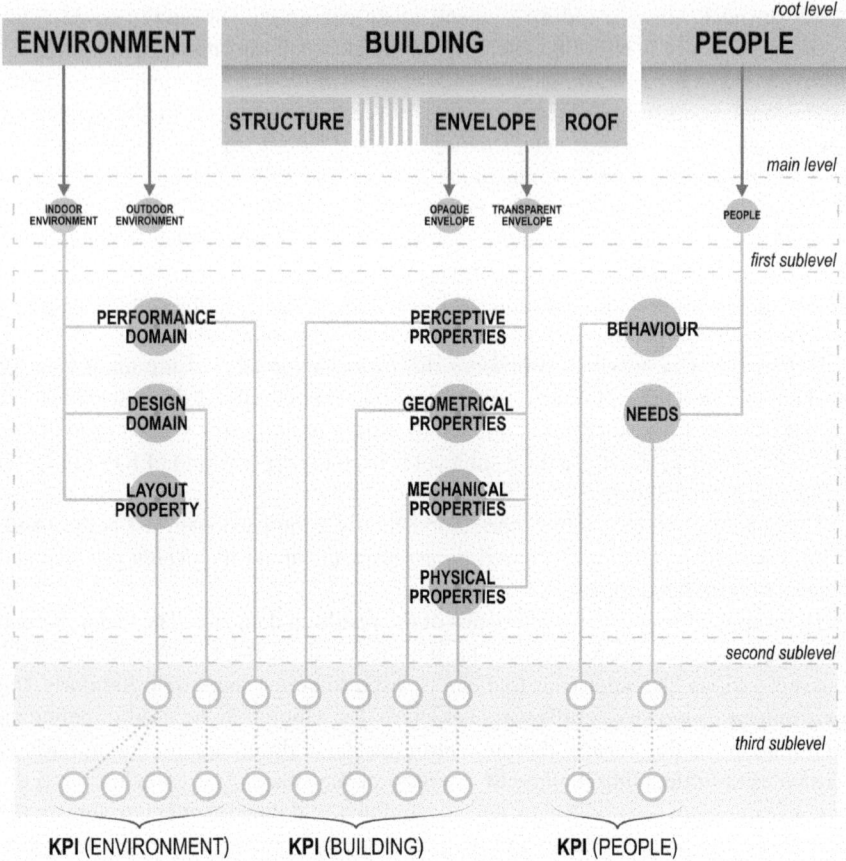

Fig. 6.3 Structure of the graph database

While no cross-group connection has been presented up to this point, as each feature of this graph is solely connected to its root path and does still not exchange any link outside of its group. The definition of cross-group connections, which essentially map the dependency network among all the recorded design factors, has been implemented into the subsequent elaboration step. However, in doing so, it is necessary to explicit the frame of reference of the dependencies that are about to be investigated.

This is far from a trivial matter. For instance, the dependency network within the design of a workspace differs significantly from that of a residential home. This difference arises from the distinct requirements, user needs, regulations, and overall functions of the spaces, all of which ultimately guide the design process. In the present context the base database has been further developed taking into account the frame of reference of a residential space.

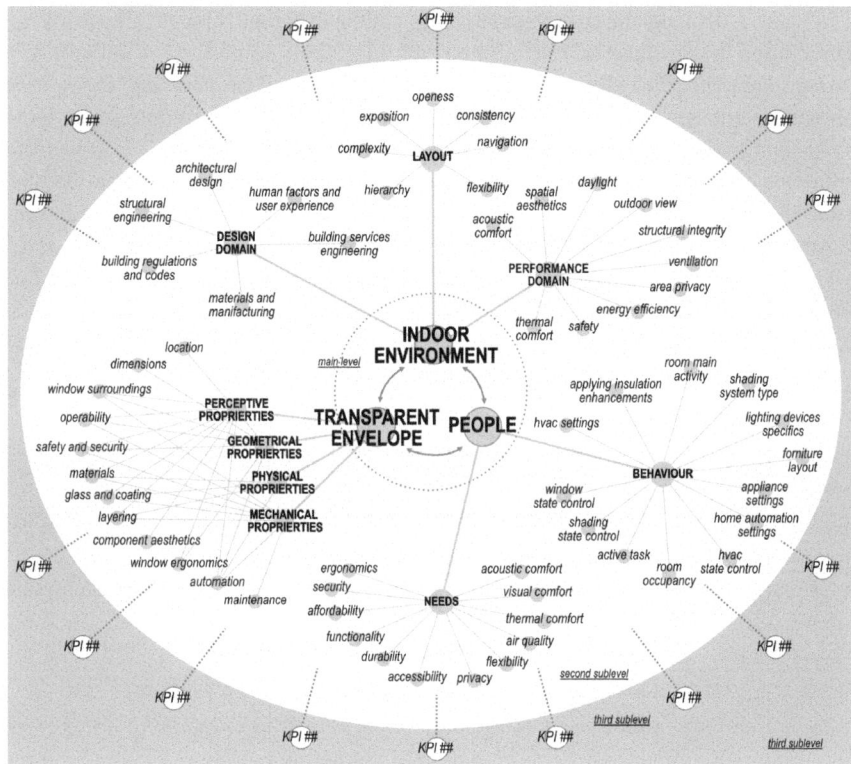

Fig. 6.4 Detailed view of the graph database entries

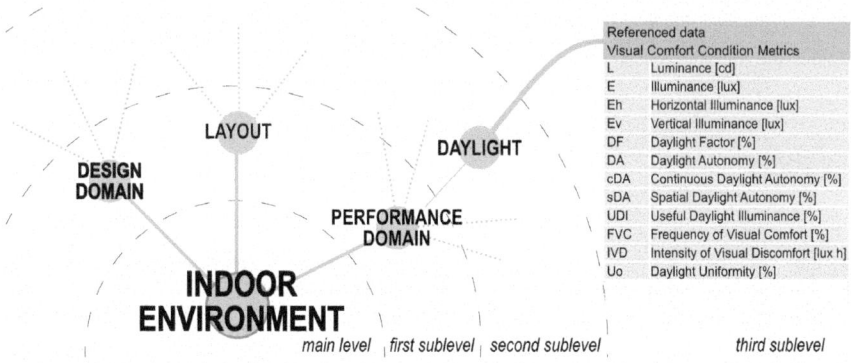

Fig. 6.5 Example of data potentially stored in the third sublevel of the database structure

Figure 6.6 displays the completed mapping of the different dependencies between the recorded design factors. Each connection represents some kind of relationship between the connected elements. The type of relationship can vary. For example, a component characteristic linked to a certain performance domain determines how any design action performed around that characteristic may impact the resulting evaluation within the linked performance domains. The resulting data can be also made available in compact form, structuring queries to extract the most relevant information in relation to specific features. In this regard, Fig. 6.7 reports the result of two simple queries, where the dependency network centred around the performance domain of daylight is isolated and highlighted. Using these queries, it is possible to monitor the potential correlation between any elements of the graph database, that is, between any elements of the design.

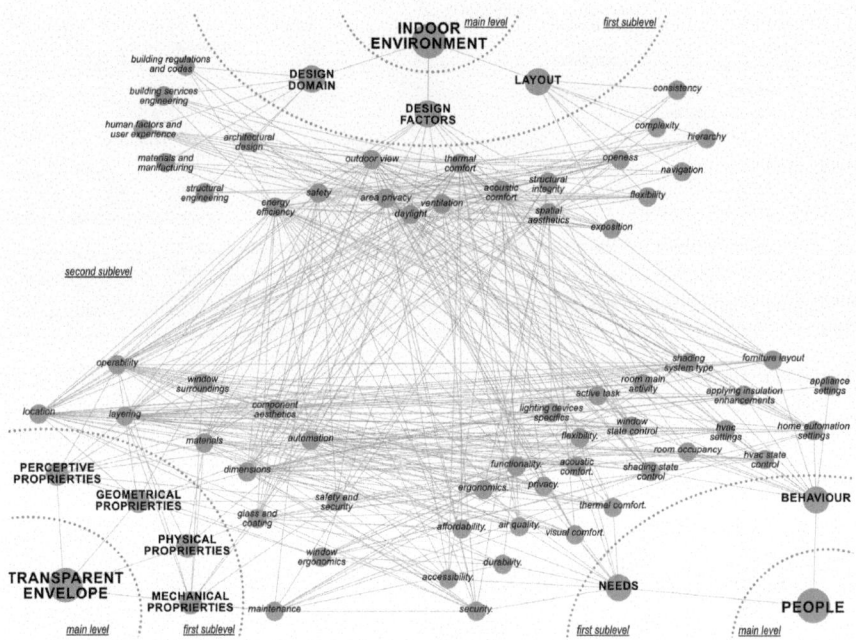

Fig. 6.6 The graph database resulting from the description of the dependency network of the different design elements once related to a housing setting

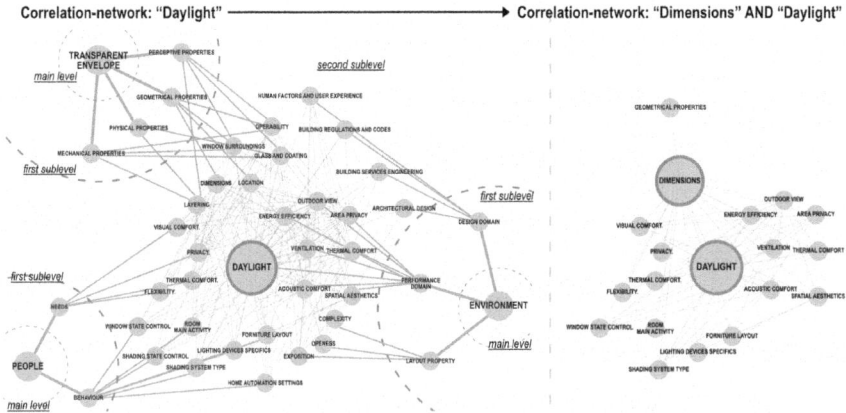

Fig. 6.7 (Left) The correlation-network of all entities directly related to "daylight". (Right) subnetwork of all entities simultaneously related to the transparent envelope property of "Dimensions" in the performance domain of "daylight"

6.3 Methodology for Text-Based Analysis and Validation of Dependency Network

Each connection within the database structure has been manually established using a double-review approach to ensure appropriate data accuracy. On one hand, empirical observations from standard design practices have been used to develop the most obvious dependencies. On the other hand, a comprehensive cross-review of existing literature has been conducted to validate the multiple possible connections. Figure 6.8 displays the workflow of the subsequent operations.

The cross-examination employed the application of text mining procedures to automate the retrieval of relevant information about the connection of specific topics across the literature and the public web spaces. Text mining is a computational process that aids in the generation of term maps from a collection of documents. A term map is a two-dimensional representation where the placement of different terms indicates their relatedness. Essentially, the closer two terms are on the map, the stronger their relationship is [11]. The concurrent use of different keywords has been recognized as a strong indicator of the relationship between various topics. Based on this premise, a series of search queries related to transparent envelope design were performed both across academic literature and the common use web search engine. The core subject of the queries search and validation procedure is here determined to be the list of transparent envelope design properties listed up to this moment (e.g., location, dimension, material, etc....). The search rules have been developed using a term matrix, as shown in Fig. 6.9. Multiple lists of search terms related to the transparent envelope are concatenated with a core group of terms that

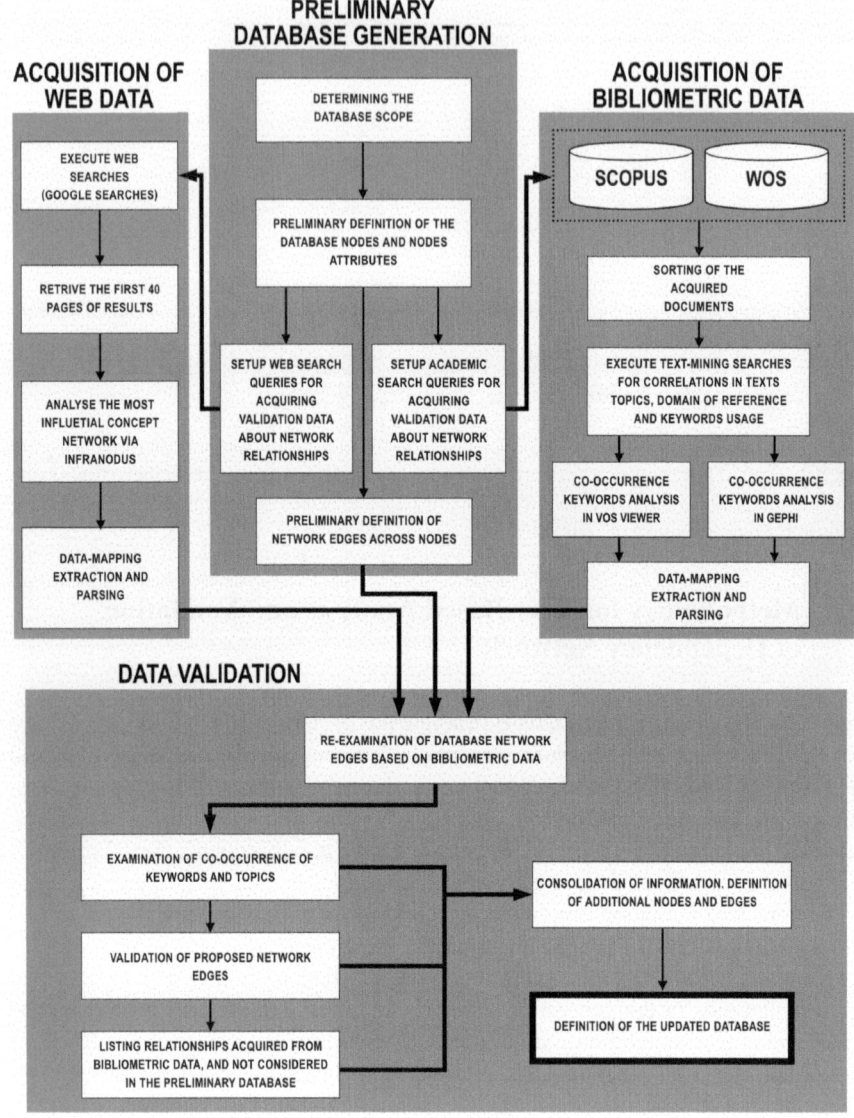

Fig. 6.8 Workflow of the cross-examination of thematic co-occurrence in the scope of transparent envelope design and performance evaluation

frame the reference for building envelope design. The complete list of variable terms is displayed in Fig. 6.10.

Public web data has also been investigated by implementing the graph-based AI-powered service infranodus [12]. The tool can retrieve a selected number of results from any web search and then automate the translation of the contents into a

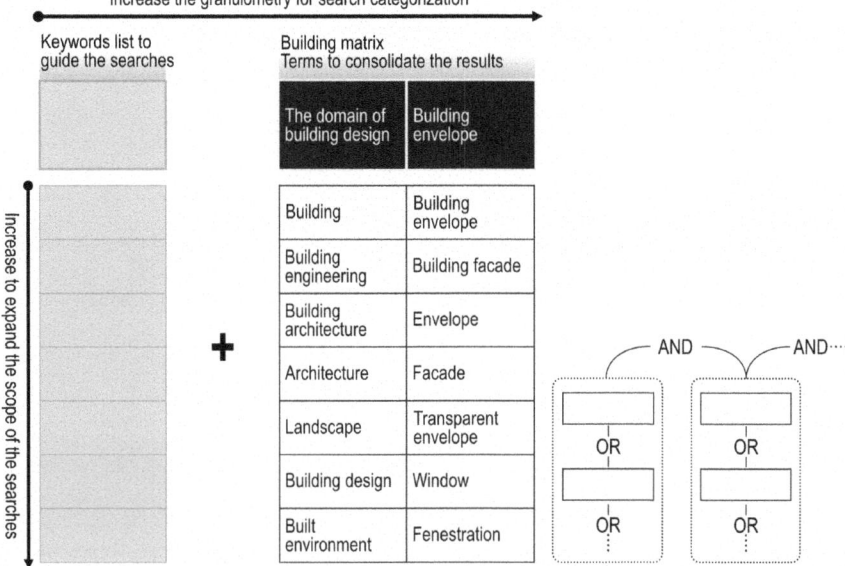

Fig. 6.9 Term matrix used to structure the necessary search queries

knowledge graph network. Essentially, it analyses multiple corpora of data to generate a corresponding graph database. This database is populated by the most recurrent words connected via edges, which in turn are determined by the concurrent use of different words in the same sentence. Since the output is a graph database, further operations can be performed to detect clusters of words with high connections or to identify the most influential elements. It can also reveal notable gaps in the dataset. The analysis of word use frequency and co-occurrence within each specific query validated previously proposed database connections. In this regard, Figs. 6.11 and 6.12 display the frequency of author keywords in the dataset from Scopus queries developed using the term matrix.

6.4 Using Graph Databases to Pre-assess the General Ranking System of Design Factors

Chapter 2, Sect. 2.2, briefly introduced the challenge of selecting and weighing specific design variables when dealing with strategic planning in the early stages of design. In this chapter, a possible tool to aid this endeavour is seen in a specifically designed graph database. This section further examines how such a generic yet flexible tool can effectively perform, starting with the necessity to establish a ranking order among design variables. In the context of a building design project, not all design elements carry equal weight. Certain aspects inevitably take precedence, due

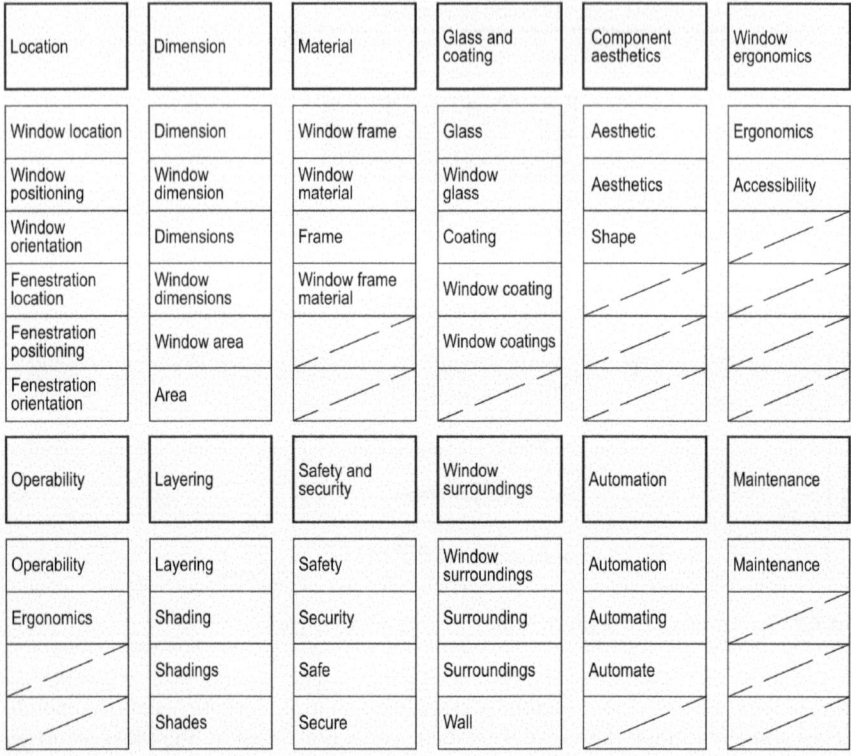

Fig. 6.10 Term lists are used to structure the necessary search queries in the scope of the many transparent envelope design properties

to specific design goals, the environment site, and user needs. Therefore, comprehending the extent of influence of these elements as soon as early-stage design is crucial to facilitate more effective and targeted decision-making processes [13]. This same question has been successfully addressed via the use of a complex sensitivity analysis setup in multiple contexts. Sensitivity analysis in the context of academic research in building design and construction engineering refers to a method used to determine how different values of an independent variable will impact a particular dependent variable under a given set of assumptions. This technique is used within specific boundaries that depend on one or more input variables, such as the impact that geometrical variations in a project may have on the energy efficiency of the overall design. Sensitivity analysis can therefore be developed to assess the impact ranking of many variables in affecting a given building's performance [14]. However, sensitivity analysis alone cannot be used to extract useful information in the form of a standardized model. This is because sensitivity analysis in building design typically is carried out on a case-by-case basis due to the numerous variables involved in each unique project. These variables can include, among others, the building's location, its intended use, the local climate, the materials used, the building's orientation, and

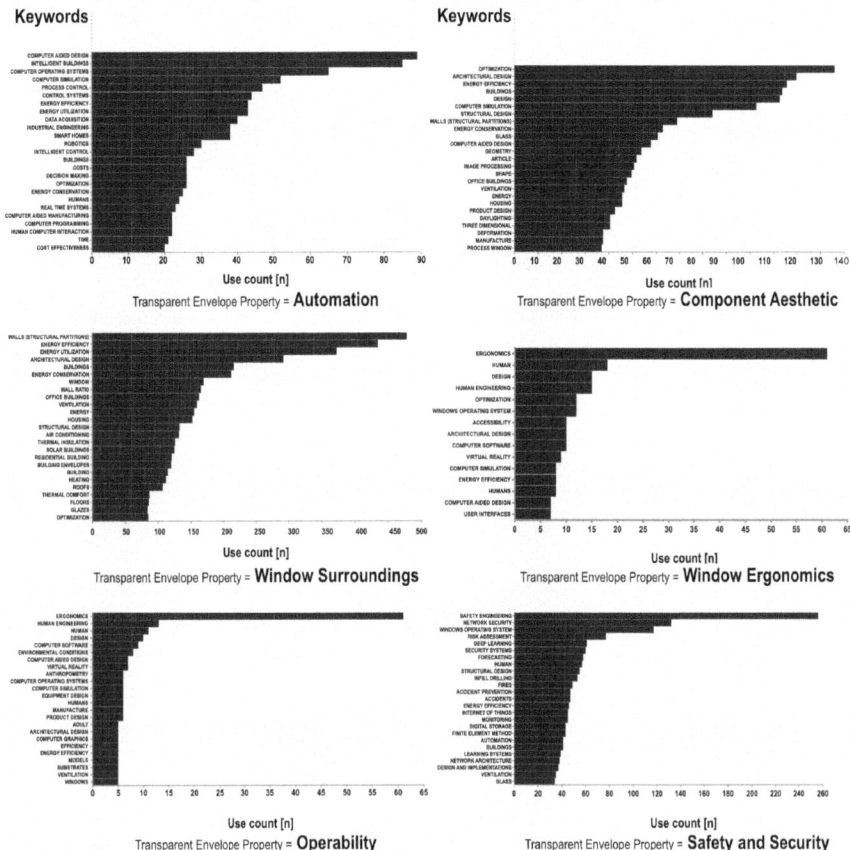

Fig. 6.11 The figure illustrates the frequency of the most commonly used author keywords across each text group, derived from search queries in academic databases. These text groups are organized based on the keywords: automation, component aesthetics, window surroundings, window ergonomics, operability, and safety and security

its architectural style. Each of these factors can significantly influence the results of a sensitivity analysis. For instance, the impact of window size on a building's energy consumption could be different in a hot, sunny climate compared to a cold, cloudy one. Therefore, while general trends and principles may be broadly identified, the specific results of a sensitivity analysis usually aren't directly transferable from one project to another. Instead, each project requires its separate analysis, shaped to its unique conditions.

However, the level of complexity addressed by sensitivity analysis may not be appropriate for the level of detail typically reached during early-stage design (see Chap. 2, Sect. 2.1). The generic nature of the model at hand, coupled with the necessity to rapidly parse a vast array of design choices, means that early-stage design does not require a precise and definite estimate of the impact of many design variables.

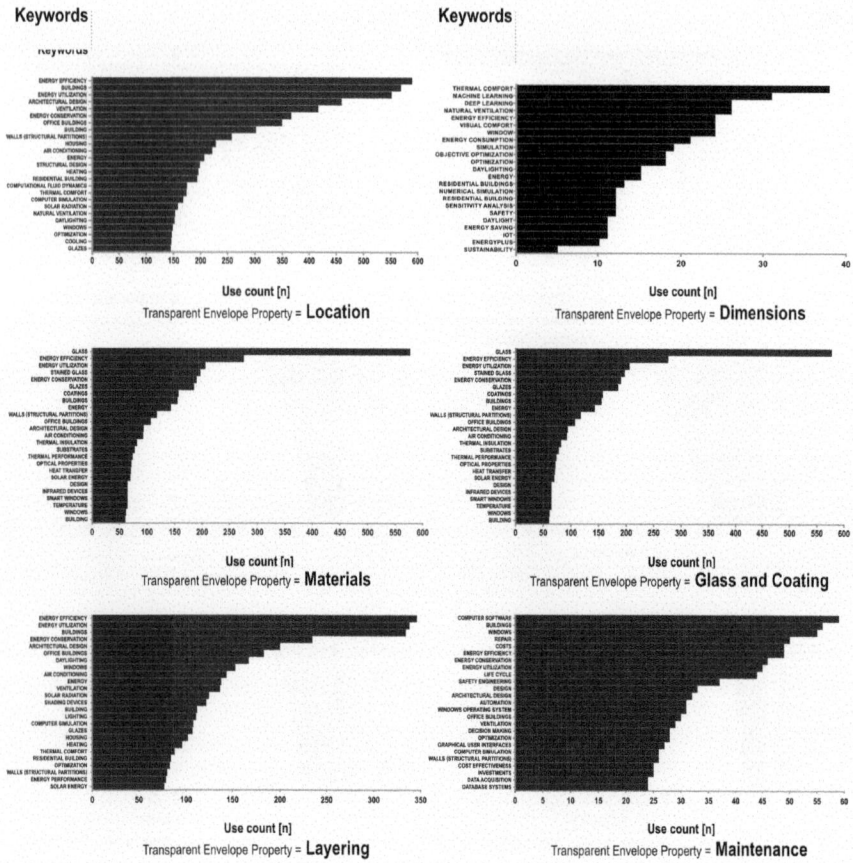

Fig. 6.12 The figure illustrates the frequency of the most commonly used author keywords across each text group, derived from search queries in academic databases. These text groups are organized based on the keywords: location, dimensions, materials, glass and coating, layering, and maintenance

Instead, it needs a global overview to iteratively construct the design strategy. Indeed, the capability of accessing a swift process to output impact rankings of multiple design elements and factors regarding a specific goal is a valuable possibility within a vast number of scenarios, not to mention AI-related applications.

To demonstrate a simulated application of the graph database data to extract a variable ranking, it is necessary first to determine a broader weight categorization per performance domains (as listed in Fig. 6.4). In this instance, two rankings are built around the concept of well-being. One scale focuses on what may be defined as a baseline level of weight, while the other prioritizes the improvement of psychological well-being.

Starting with the baseline ranking, a fundamental observation can be attributed to James Marston Fitch who noted how aesthetic judgments become possible only as long as a person is not suffering overly stressful levels of discomfort (i.e. thermal, acoustic and luminous discomfort). The level of contentment or dissatisfaction a person experiences in a particular situation is closely linked to their physical well-being. When there is an optimal level of sensory stimulation, neither excessive nor insufficient, it allows for the most effective use of one's analytical abilities to evaluate and engage with the situation or any of its aspects [1]. This would suggest that performance domains oriented to evaluating different conditions for reaching optimal physiological well-being should be implemented with higher priority than other factors, as achieving positive results on these aspects determines a sort of foundational baseline for a higher form of psychological well-being. In this regard, an even greater priority should be given to the overall structural integrity of the habitable environment, as safety is recognized as a core need, second only to essential physiological requirements like food and shelter. Consequently, a sense of safety can be seen as a prerequisite to fulfil advanced emotional needs, forming an integral part of overall wellness [15, 16].

In the context of the building's transparent envelope design, what may be defined as "aesthetic judgment" can assume many forms, ranging from the architectural design of the overall transparent surfaces distributions and shapes, their integration with space, the outdoor view, or other features such as automation or ergonomics [17–20].

Regarding the outdoor view, or window view [21], multiple positive effects on people's psychological well-being have been the focus of numerous recent studies within the field of built environment research. Among the correlated findings, multiple independent studies positively regarded the availability of a direct visual connection to nature. This concept, often referred as "Biophilia", or "Biophilia hypothesis", advocates that the experience of nature has a positive influence on the well-being and satisfaction of people [22]. This is particularly relevant within closed habitable spaces [23], where it has been reported to decrease discomfort and stress, thereby mitigating the adverse effects of performing activities with a high cognitive load, such as job-related tasks [24–26]. The degree of such improvement in the overall psychological benefits reaches the point to exhibit sensible impacts on a reduction in the likelihood of employees quitting their jobs [27]. It also increases the appeal and productivity of workspaces with appropriate visual characteristics [28].

Exposure to outside views through the building's transparent envelope can also affect the psychological well-being of occupants, especially in residential environments, via its inversed action. As noted by Zheng et al. the visual component determined by the "strangers' ability to view in", often defined as "privacy", has been limitedly addressed in past decades, although it has now become a major concern in evaluating environmental quality [29]. On this topic, Newell reviewed that although the concept of privacy is complex and multifaceted, varying in scope and definition depending on different circumstances (e.g., cultural background, field of research, environment typology, etc....), it is possible to outline clear situations involved with privacy issues by comparing the many specific definitions of the concept [30]. Among these situations, the "spatial privacy invasion via visual access" is one of the several ways in which privacy may be invaded, and it is particularly relevant in window and building envelope design [30]. Due to this, privacy can be also defined solely in terms of visual privacy, which refers to the capacity to engage in everyday tasks at one's residence without the scrutiny of outsiders, which includes neighbours and individuals passing by [31].

Visual privacy is therefore another type of exchange mediated by the technological component of the transparent envelope, and similarly to the outdoor view, it impacts different aspects of the inhabitant's well-being on a psychological level [31, 32].

This brief review of the performance-oriented domains mapped in the graph database up to this point, can be used to guide the definition of a preliminary ranking order.

In relation to the target of addressing the occupant's well-being two separate evaluations could be outlined. A first evaluation can approach the target of well-being prioritizing the necessary conditions to ultimately favour the actualization of the proper environmental conditions to experience the many forms of psychological well-being. This frame of reference results in assigning the top priority to safety-oriented domains, subsequently placing the many performance domains oriented towards physiological well-being, and lastly the ones related to psychological well-being. Table 6.1 displays this ranking distribution. The present approach could be a consequence of targeting with increased emphasis and certainty the base conditions for the safety and physical comfort of a building. On the other hand, it could be also applied to an opposite target, which explicitly prioritizes all the performance domains related to psychological well-being. In such a case, the ranking score could be flipped, as Table 6.2 shows, due to the desire to explicitly increase the capabilities of a design to channel the right characteristic for high-quality spaces.

Table 6.1 General weight scores associated with the database links based on the criteria of focusing on the base conditions to develop a satisfactory environment to address physiological forms for well-being

General weight link	Performance domains	Well-being category
3.0	Structural integrity, safety	Baseline safety conditions
2.0	Daylight, thermal comfort, energy efficiency, acoustic comfort, ventilation	Physiological well-being
1.0	Outdoor view, area privacy, spatial aesthetics	Psychological well-being

Table 6.2 General weight scores associated with the database links based on the criteria of focusing directly on the performance domains involved with psychological forms for well-being

General weight link	Performance domains	Well-being category
3.0	Outdoor view, area privacy, spatial aesthetics	Psychological well-being
2.0	Daylight, thermal comfort, energy efficiency, acoustic comfort, ventilation	Physiological well-being
1.0	Structural integrity, safety	Baseline safety conditions

6.5   Implementing Weighted Graph Databases for Efficient Retrieval of Ranked Information

The next point presented in this context is the use of the resulting dependency network to extract weighted sub-networks. The additional data about the general weighted priority of the mapped performance domains, in fact, allows for the implementation of such weighted links to rank other elements of the database, based on their number of connections to performance domains. This type of query is easily performed within graph databases via the use of specific evaluation metrics called degree [33]. The degree of a node refers to the total number of connections or edges it possesses, regardless of whether these are incoming or outgoing. Essentially, it's the cumulative count of a node's connections [34, 35]. Moreover, if edges feature a weighted rank, this can be accounted by calculating the weighted degree of a node. The weighted degree is similar to the degree; it's determined by the number of edges a node has. However, it also considers the weight associated with each edge, thus offering a more nuanced measure [34, 36]. In this case, as each performance domain has been weighted with a specific rank, it is possible to extract the weighted degree of other connected elements to generate simple ranking of impacts regarding the proposed design goal of well-being within the window design process. However, it's important to acknowledge that the criteria used to develop the weight values, while effective for establishing a preliminary understanding, fall short of capturing the complexities inherent in the design process. However, despite their simplicity, these results may provide valuable insights for early-stage design, offering a generalized, understanding of how certain design decisions may influence the building performances and well-being. In particular, Figs. 6.13 and 6.14 display the results of implementing the ranking determined by targeting the necessary conditions of leverage physiological well-being (the baseline scale), while Figs. 6.15 and 6.16 display the results of targeting explicitly physiological well-being.

6.6   Trade-Offs Among Performance Domains: A Mapping Exercise Using Graph Databases

To conclude the definition of the fundamental structure of the presented database, it is proposed one last aspect to improve within the present graph database. In this regard, it is thought that the database should not stop at considering and storing only the relative dependence of performance domains on various design factors, but also the interdependencies among the performance domains themselves. In fact, as the graph database has revealed, different design factors can simultaneously impact multiple performance domains, highlighting the interconnected nature of all these building aspects. However, these connections are not neutral. Depending on the circumstances, they can manifest as either synergistic or antagonistic relationships. To put it simply, in the context of guiding decision-making strategies within early-stage design, it's

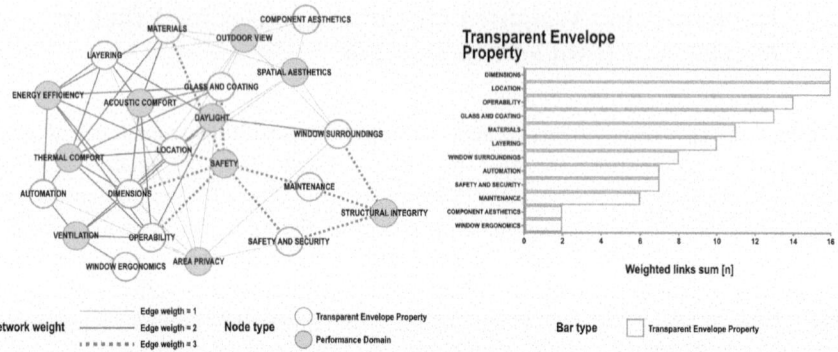

Fig. 6.13 (Left) Highlight Sub-network composed by performance domains and transparent envelope properties. The edges' thickness is proportional to their weights. Based on each property connection towards performance domains, a weighted degree is presented and ranked. (Right) Histogram based on the weighted degree scores. The diagram shows how "dimension", "location" and "operability" are the most impactful properties regarding well-being

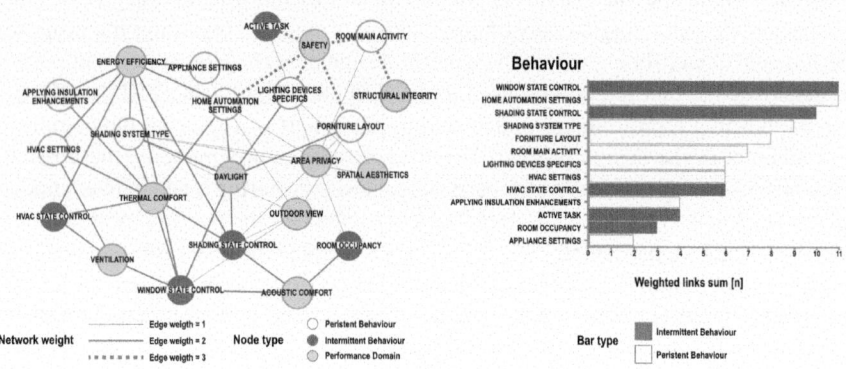

Fig. 6.14 (Left) Highlight Sub-network composed by performance domains and people behaviours. The edges' thickness is proportional to their weights. Similarly, to the previous image, based on each behaviour connection towards performance domains, a weighted degree is presented and ranked. (Right) Histogram based on the weighted degree scores. The diagram shows how "window state control", "home automation settings" and "shading state control" are the most impactful behaviours regarding well-being

not only important to have easy access to the list of possible variables to alter and understand what aspects of the design they may change and the potential impact, but also to understand how such interconnected aspects may be altered.

As Fig. 6.8 summarizes, the major issue with window design is the number of different exchanges which windows allow between the interior and exterior environments. As these exchanges happen simultaneously, filtering advantageous interactions while limiting disadvantageous ones, requires balancing the overall exchanges into an ideal equilibrium. When a type of exchange can benefit from the optimization

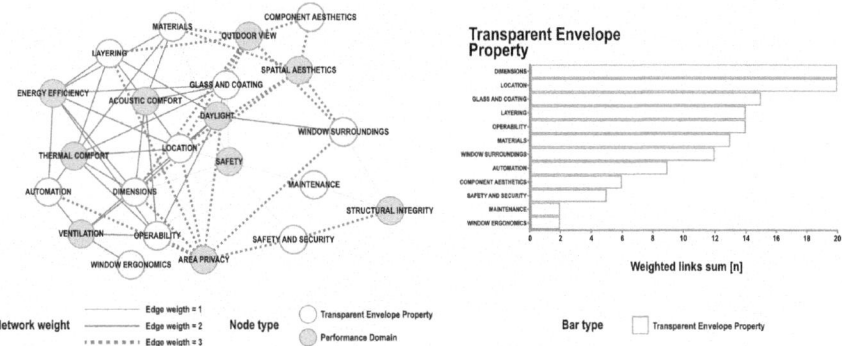

Fig. 6.15 (Left) Highlight Sub-network composed by performance domains and transparent envelope properties implementing the scores derived from Table 5. (Right) Histogram based on the weighted degree scores. The diagram illustrates how "glass and coating" and "layering" are now considered to be of increased importance

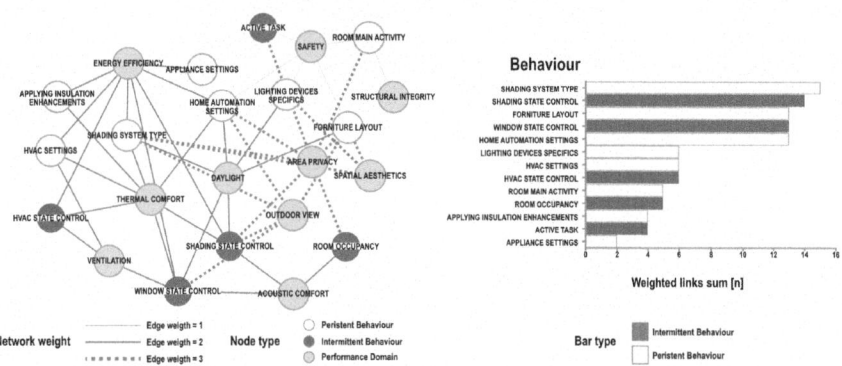

Fig. 6.16 (Left) Highlight Sub-network composed by performance domains and people behaviours implementing the scores derived from Table 5. (Right) Histogram based on the weighted degree scores. The diagram shows how "shading system type" and "shading state control" are now considered to be of increased importance

of another, it could be said that they share a synergic relationship, on the contrary, when the optimization of one exchange causes a detriment in the performance of others, it could be said that they are linked in an antagonistic relationship.

Figure 6.17 compiles a generalised relationship-matrix about the state of relationships among the mapped performance domains, reflecting the body of literature on envelope and window design that was analyzed during the study, with a focus on the performance domains of daylight and outdoor view.

The passage of light for example can affect simultaneously multiple performance domains and perceptions. Light is what allows visual perceptions, both in the form of the availability to see outside from the inside (outdoor view), from the outside to inside (view in), and within the interior environment (daylight). All these interactions

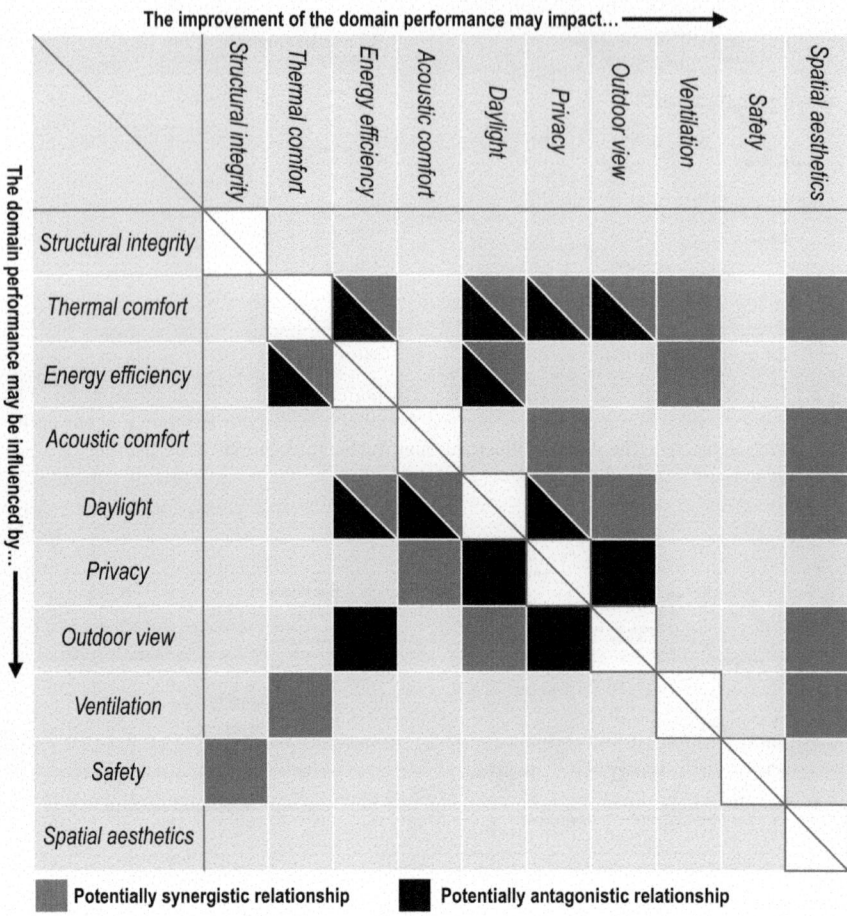

Fig. 6.17 The relationship matrix highlights the relational network between the analysed performance domains

are therefore fundamentally linked [37]. In addition, daylight possesses not only illuminating properties but also those of radiation. This results in inevitable solar heat gain, which may pose a challenge, this is because if the heat transferred to the interior environment reaches excessive levels, it then becomes a cooling load that requires removal by the air-conditioning system [38]. This statement isn't always valid, as the relationship between daylight and internal thermal control also depends on the environmental context. Therefore, based on the season and site, solar gains promoted by daylight may either decrease or increase the energy costs for temperature control. This makes the relationship between these domains both synergic and antagonistic depending on the circumstances [39].

The information added to the databases can ultimately be used to aid decision-making strategies in early-stage design. This data may set up optimization procedures that automatically establish default fitness goals to minimize or maximize multiple related performance evaluations [40, 41]. While these suggested parameters may be revised by the final user depending on the unique traits of a project and its design, they could still pose an effective aid in planning the overall design strategies to implement.

6.7 Assessing the Use of Graph Database Queries Within Simulated Design Workflows

After detailing the structure and contents of the proposed graph database, the next step is to explore its practical application in early-stage building design through simulated workflows. This exploration helps illustrate the database's potential as a decision-making aid, while also uncovering opportunities for further refinement and expansion. By understanding various design scenarios and needs, enhancements to the database can be identified, ensuring its ongoing relevance and utility in the dynamic field of architectural design.

These workflows were specifically designed to align with what are considered the primary types of design processes: preliminary planning, iterative optimization, predictive problem-solving, and a comparative approach which focuses on comparing multiple selected alternatives.

In this context, Fig. 6.18 details a possible applicative example focused on a specific set of design feature. Assuming that a specific design instance proposed within the context of early-stage design may have been evaluated as potentially energy inefficient, this should prompt a new iteration within the design stage to address this issue by applying adequate changes to the proposed design. In this regard, implementing the list of design elements collected into a sub-network focused on the building energy efficiency performance domain, the most fitting factors may be selected to guide an editing process into the design.

In conclusion, the work detailed in this chapter addresses the critical phase of early-stage design, emphasizing the importance of managing the building envelope as an effective variable to enhance a building's overall value. By focusing on the quality of space and the concerns and well-being of people, it becomes evident that laying down an effective design strategy from the start is essential to ensure the correct generation of habitable spaces. Managing complex relationships between various parts of the design serves this purpose, unlocking the capability to access, share, and implement a multi-domain overview of the intricate conceptual network of a building. The approach founded on visually interactive networks not only supports informed decision-making but also ensures that the design process remains adaptive and responsive to the diverse needs and performance targets identified in the early stages.

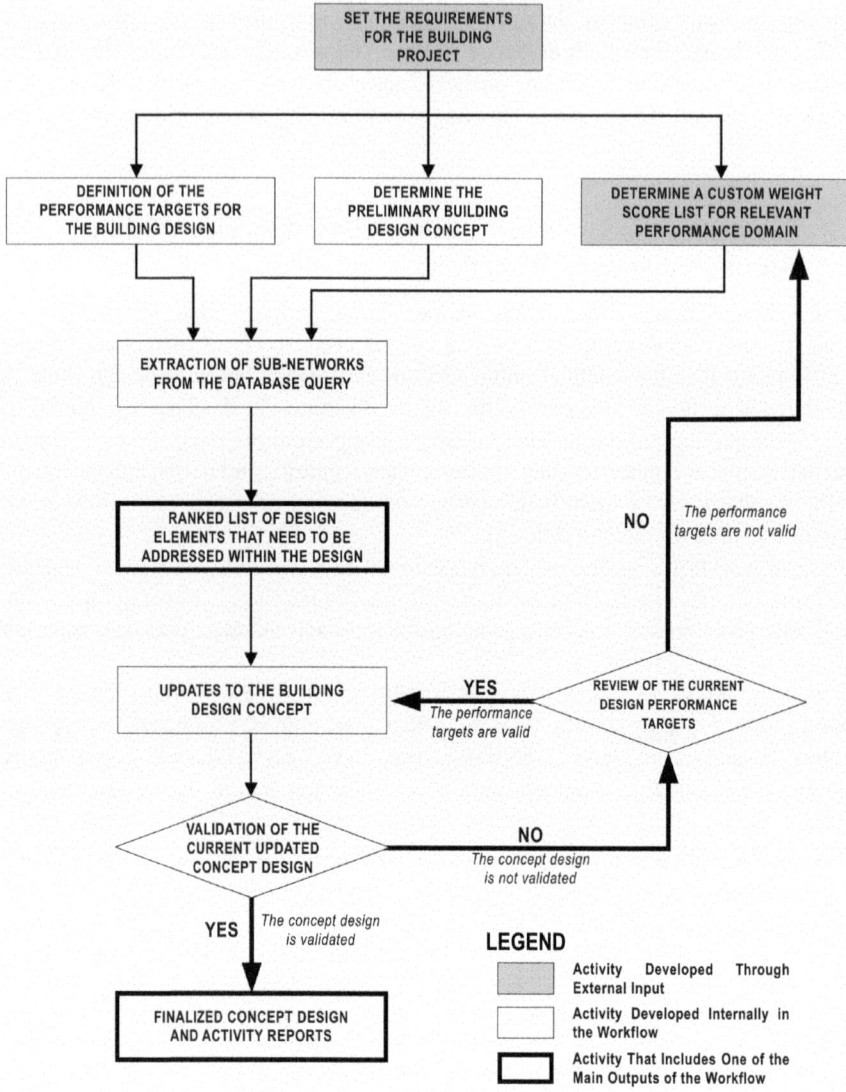

Fig. 6.18 Simulated workflow displaying the possible implementation of the graph database inside a typical optimization workflow

References

1. Fitch JM (2006) The aesthetics of function. Ann N Y Acad Sci 128:706–714. https://doi.org/10.1111/j.1749-6632.1965.tb11687.x
2. Pamungkas A (2020) Infographics and data visualizations. https://medium.com/@anggerpam ungkas/infographics-and-data-visualizations-9f5726452027. Accessed 18 Sep 2023

3. Joyce KE (2021) Graph database vs. relational database: key differences. https://www.techta rget.com/searchdatamanagement/feature/Graph-database-vs-relational-database-Key-differ ences. Accessed 23 Apr 2023
4. Robinson I, Webber J, Eifrem E (2013) Graph databases, 2nd edn. O'Reilly Media, Inc
5. Almabdy S (2018) Comparative analysis of relational and graph databases for social networks. In: 2018 1st international conference on computer applications & information security (ICCAIS). IEEE, pp 1–4
6. Jouili S, Vansteenberghe V (2013) An empirical comparison of graph databases. In: 2013 international conference on social computing. IEEE, pp 708–715
7. Roughan M, Tuke J (2015) Unravelling graph-exchange file formats. https://doi.org/10.48550/arXiv.1503.02781
8. The Open Graph Viz Platform. https://gephi.org/. Accessed 23 Apr 2023
9. Nardi G (1977) Progettazione architettonica per sistemi e componenti, 2nd edn. Franco Angeli
10. Rodríguez Iturriaga M (2021) Learning from COVID-19: the role of architecture in the experience of urban landscapes. Ri-Vista Research for landscape architecture 19:122–137. https://doi.org/10.36253/rv-10182
11. van Eck NJ, Waltman L (2011) Text mining and visualization using VOSviewer
12. Infranodus. https://infranodus.com/. Accessed 19 Aug 2023
13. Østergård T, Jensen RL, Maagaard SE (2017) Early building design: informed decision-making by exploring multidimensional design space using sensitivity analysis. Energy Build 142:8–22. https://doi.org/10.1016/j.enbuild.2017.02.059
14. Heiselberg P, Brohus H, Hesselholt A et al (2009) Application of sensitivity analysis in design of sustainable buildings. Renew Energy 34:2030–2036. https://doi.org/10.1016/j.renene.2009.02.016
15. Zeng E, Dong Y, Yan L, Lin A (2022) Perceived safety in the neighborhood: exploring the role of built environment, social factors, physical activity and multiple pathways of influence. Buildings 13:2. https://doi.org/10.3390/buildings13010002
16. Mouratidis K (2019) The impact of urban tree cover on perceived safety. Urban For Urban Green 44:126434. https://doi.org/10.1016/j.ufug.2019.126434
17. Kaplan S (1987) Aesthetics, affect, and cognition. Environ Behav 19:3–32. https://doi.org/10.1177/0013916587191001
18. Vartanian O, Navarrete G, Chatterjee A et al (2013) Impact of contour on aesthetic judgments and approach-avoidance decisions in architecture. Proc Natl Acad Sci 110:10446–10453. https://doi.org/10.1073/pnas.1301227110
19. Toromanoff A (2021) Curved: Bending architecture, 1st edn. Lannoo Publishers
20. Montjoy V (2022) The comeback of curved design: materials that can bend and curl. https://www.archdaily.com/977895/the-comeback-of-curved-design-materials-that-can-bend-and-curl?ad_source=search&ad_medium=search_result_articles. Accessed 29 May 2023
21. Ko WH, Schiavon S, Altomonte S et al (2022) Window view quality: why it matters and what we should do. LEUKOS 18:259–267. https://doi.org/10.1080/15502724.2022.2055428
22. Wilson EO (1984) Biophilia, 3rd edn. Harvard University Press
23. Stouhi D (2023) Biophilic interiors: 21 projects that blend architecture with nature. https://www.archdaily.com/995875/biophilic-interiors-21-projects-that-blend-architecture-with-nat ure?ad_source=search&ad_medium=search_result_articles. Accessed 9 May 2023
24. Joye Y (2007) Architectural lessons from environmental psychology: the case of biophilic architecture. Rev Gen Psychol 11:305–328. https://doi.org/10.1037/1089-2680.11.4.305
25. Ko WH, Schiavon S, Zhang H et al (2020) The impact of a view from a window on thermal comfort, emotion, and cognitive performance. Build Environ 175:106779. https://doi.org/10.1016/j.buildenv.2020.106779
26. Kellert SR (2012) Building for life: designing and understanding the human-nature connection, 1st edn. Island press
27. Leather P, Pyrgas M, Beale D, Lawrence C (1998) Windows in the workplace. Environ Behav 30:739–762. https://doi.org/10.1177/001391659803000601

28. Altomonte S, Allen J, Bluyssen PM et al (2020) Ten questions concerning well-being in the built environment. Build Environ 180:106949. https://doi.org/10.1016/j.buildenv.2020.106949

29. Zheng H, Wu B, Wei H et al (2021) A quantitative method for evaluation of visual privacy in residential environments. Buildings 11:272. https://doi.org/10.3390/buildings11070272

30. Newell PB (1995) Perspectives on privacy. J Environ Psychol 15:87–104. https://doi.org/10.1016/0272-4944(95)90018-7

31. Al-Kodmany K (1999) Residential visual privacy: traditional and modern architecture and urban design. J Urban Des (Abingdon) 4:283–311. https://doi.org/10.1080/13574809908724452

32. de Macedo PF, Ornstein SW, Elali GA (2022) Privacy and housing: research perspectives based on a systematic literature review. J Hous Built Environ 37:653–683. https://doi.org/10.1007/s10901-022-09939-z

33. (2010) Gephi tutorial quick start. https://gephi.org/tutorials/gephi-tutorial-quick_start.pdf. Accessed 12 Jun 2023

34. Bhasin J (2019) Graph analytics—introduction and concepts of centrality. https://towardsdatascience.com/graph-analytics-introduction-and-concepts-of-centrality-8f5543b55de3. Accessed 12 Jun 2023

35. Heymann S (2015) Gephi wiki degree. https://github.com/gephi/gephi/wiki/Degree. Accessed 12 Jun 2023

36. Weighted Degree Centrality (2023). https://docs.tigergraph.com/graph-ml/current/centrality-algorithms/weighted-degree-centrality. Accessed 12 Jun 2023

37. Jamrozik A, Clements N, Hasan SS et al (2019) Access to daylight and view in an office improves cognitive performance and satisfaction and reduces eyestrain: a controlled crossover study. Build Environ 165:106379. https://doi.org/10.1016/j.buildenv.2019.106379

38. Futrell BJ, Ozelkan EC, Brentrup D (2015) Optimizing complex building design for annual daylighting performance and evaluation of optimization algorithms. Energy Build 92:234–245. https://doi.org/10.1016/j.enbuild.2015.01.017

39. Chi DA, Moreno D, Navarro J (2018) Correlating daylight availability metric with lighting, heating and cooling energy consumptions. Build Environ 132:170–180. https://doi.org/10.1016/j.buildenv.2018.01.048

40. Rutten D (2010) Evolutionary principles applied to problem solving. https://www.grasshopper3d.com/profiles/blogs/evolutionary-principles. Accessed 13 Jun 2023

41. Ciardiello A, Rosso F, Dell'Olmo J et al (2020) Multi-objective approach to the optimization of shape and envelope in building energy design. Appl Energy 280:115984. https://doi.org/10.1016/j.apenergy.2020.115984

Chapter 7
Unlocking the Future: Reaching a Conclusion

Abstract This study examined the complexities of building envelope design, focusing on its multidimensional challenges and the need to balance a broad range of inter-connected factors. Central to the research is the concept of "unlocking", which refers to overcoming previous limitations through innovative strategies. The work explores a spectrum of current innovative trends within the framework of building envelope design, ranging from the macro-scale to the micro-scale. The results emphasize the shift from a "one-size-fits-all" approach to adaptive, human-centered design, highlighting increasingly precise methods for addressing occupants' needs. The study concludes that successful design solutions require interdisciplinary collaboration, and that the proficient implementation of innovative solutions plays a key role in creating sustainable and resilient building envelopes that meet both current and future needs.

Throughout the preceding chapters, the complex topic of building envelope design has been dissected along various analytical dimensions. These dimensions were carefully selected to address the many challenges that contemporary building envelope design faces, ensuring a comprehensive exploration of practical and theoretical implications. In this regard, a fundamental concept recurrently emphasized in the presented body of research is the idea that building envelope design represents a multifaceted challenge, whose solution requires a comprehensive understanding of various interconnected factors. This is because building design has evolved into a complex, multi-criteria problem where numerous aspects such as safety, sustainability, comfort, and well-being must be precisely balanced. The multitude of variables places greater responsibility on the designer, adding complexity to their role. However, this challenge is aided by the support of various technological innovations, which help navigate these demands and enable the development of a clear strategic vision.

This fundamental view on the topic has guided the choice of projects for presentation, with each project ultimately charting a path for innovation in building envelope design.

A. G. Mainini et al., *Unlocking the Potential of Building Envelopes*,
PoliMI SpringerBriefs, https://doi.org/10.1007/978-3-031-75298-8_7

Different aspects interlinked to building envelope design, ranging from innovative advanced materials and complex strategies for decision-making approaches in building design, to environmental quality control technologies, servitization for energy production, and the design towards visual comfort, all converge to a common nexus defined in the text as "unlocking".

Unlocking, in its literal sense, involves gaining an advantage by rendering accessible a sealed obstruction. In the realm of building envelope design, this concept can refer to the development of envelopes with enhanced capabilities using design strategies that, for various reasons, might have been normally unapproachable. Depending on the context, this unlock can widely vary in appearance: novel materials could unlock the capability to design complex shapes, while product transfers from different sectors could represent an opportunity to develop more performative structures at decreased costs. In all cases, however, "unlock" represents the same concept: achieving enhanced results by identifying and applying novel concepts previously unavailable, characterizing innovation in its purest form.

However, the pursuit of unlocking is not merely a goal in itself; instead, it serves as a strategic tool to seek balance amid the ever-growing intricacy of contemporary design, finding the means to implement actions that can holistically provide optimal solutions to interconnected design challenges.

The process of unlocking is therefore achieved through innovative strategies that span from macro-scale to micro-scale, approaching not only the design of the building envelope as a system but also its constituent elements, up to the very materials used to develop specific building components. Moreover, the unlocking process also involves a reverse engineering approach that, starting from the impact (social, economic, environmental), traces back to the best strategy/solution among the many possibilities (a sort of technology foresight applied to the building envelope). Advanced and innovative materials play a crucial role in addressing and solving these challenges in building design. Smart materials, which can undergo reversible changes in response to environmental stimuli, enable building envelopes to dynamically adjust their performance. These materials contribute to the creation of responsive structures that enhance energy efficiency, improve indoor comfort, and reduce the reliance on mechanical systems for heating, cooling, and lighting. Building envelopes, traditionally seen as mere separators between indoor and outdoor spaces, now play a pivotal role as dynamic filters. They have reached the point where they connect private and public spaces, providing personalized functions to each side and adding another layer of complexity to the design process.

As building design objectives have evolved, the focus has shifted from merely supporting safety to delivering comfort and ultimately enhancing the well-being and health of occupants. In this regard, to effectively provide comfort, building designs can no longer adhere to a "one-size-fits-all" approach. Instead, they must dynamically address individual preferences in a human-centric manner. This necessitates continuous, rather than periodic, acknowledgement of occupant needs, facilitated by advanced data monitoring and dynamic feedback systems. The goal is to create ever-adaptive spaces that respond in real-time to the changing needs of their users.

However, the myriad aspects of building design and use do not always align seamlessly. Addressing one aspect may lead to a decrease in efficiency in another. For instance, prioritizing thermal comfort might impact energy efficiency or aesthetic quality. These design factors are framed within a hierarchy of importance, with specific rank orders and trade-offs between them. Navigating this complexity requires a multi-disciplinary approach, where parametric design and optimization strategies play a crucial role. These methodologies help determine the equilibrium state among all contributing factors, ensuring a balanced and integrated design solution. A sustainable building should be conceived as the sum of three aspects: a sustainable life cycle, an optimal urban environment, and exceptional indoor quality. Therefore, the aesthetic experience of space, well-being, and sustainable practices must be interconnected. This holistic approach ensures that buildings not only meet contemporary needs but also preserve resources for future generations. At the crossroads of further innovation lies data and technology, particularly the Internet of Things (IoT), Artificial Intelligence (AI), and advanced computational methods. These technologies hold immense potential for revolutionizing building envelope design by enabling predictive modelling, real-time monitoring, and adaptive responses. However, true innovation can only occur if there is be concerted effort to integrate these technologies into the design process holistically. This integration must be supported by a collaborative framework that brings together engineers, architects, data scientists, and other stakeholders, ensuring that diverse expertise contributes to the creation of more resilient, efficient, and human-centric building envelopes. By continuously monitoring and adapting to the needs of occupants, leveraging parametric design and optimization, and fostering interdisciplinary collaboration, we can create building envelopes that are not only efficient and resilient but also enhance the well-being of their users. The journey towards more intelligent and responsive building envelopes is ongoing, requiring a sustained commitment from all stakeholders to realize the transformative potential of this field. Through the thoughtful integration of advanced materials, continuous data monitoring, and dynamic feedback, we can achieve the harmonious balance necessary for creating buildings that truly enhance the quality of life and the environment.

GPSR Compliance

The European Union's (EU) General Product Safety Regulation (GPSR) is a set of rules that requires consumer products to be safe and our obligations to ensure this.

If you have any concerns about our products, you can contact us on ProductSafety@springernature.com

In case Publisher is established outside the EU, the EU authorized representative is:

Springer Nature Customer Service Center GmbH
Europaplatz 3
69115 Heidelberg, Germany

The manufacturer's authorised representative in the EU is Springer
Nature Customer Service Centre GmbH, Europaplatz 3, 69115 Heidelberg,
Germany. If you have any concerns regarding our products, please
contact ProductSafety@springernature.com

Printed and bound by CPI Group (UK) Ltd, Croydon, CR0 4YY

29/04/2026

02099547-0001